MW00612764

Energy Efficiency and Sustainable Consumption

Energy, Climate and the Environment Series

Series Editor: David Elliott, Professor of Technology, Open University, UK

Titles include:

David Elliott *(editor)*
NUCLEAR OR NOT?
Does Nuclear Power Have a Place in a Sustainable Future?

David Elliott *(editor)*
SUSTAINABLE ENERGY
Opportunities and Limitations

Horace Herring and Steve Sorrell *(editors)*
ENERGY EFFICIENCY AND SUSTAINABLE CONSUMPTION
The Rebound Effect

Catherine Mitchell
THE POLITICAL ECONOMY OF SUSTAINABLE ENERGY

Joseph Szarka
WIND POWER IN EUROPE
Politics, Business and Society

Energy, Climate and the Environment

Series Standing Order ISBN 0–2300–0800–3

You can receive future titles in this series as they are published by placing a standing order. Please contact your bookseller or, in case of difficulty, write to us at the address below with your name and address, the title of the series and the ISBN quoted above.

Customer Services Department, Macmillan Distribution Ltd, Houndmills, Basingstoke, Hampshire RG21 6XS, England

---

Energy Efficiency and Sustainable Consumption

The Rebound Effect

Edited by

Horace Herring
Visiting Research Fellow,
Energy and Environment Research Unit
The Open University, UK

and

Steve Sorrell
Senior Fellow, Sussex Energy Group
University of Sussex, UK

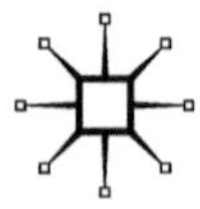

First published 2009 by
PALGRAVE MACMILLAN

Palgrave Macmillan in the UK is an imprint of Macmillan Publishers Limited, registered in England, company number 785998, of Houndmills, Basingstoke, Hampshire RG21 6XS.

Palgrave Macmillan in the US is a division of St Martin's Press LLC, 175 Fifth Avenue, New York, NY 10010.

Palgrave Macmillan is the global academic imprint of the above companies and has companies and representatives throughout the world.

Palgrave® and Macmillan® are registered trademarks in the United States, the United Kingdom, Europe and other countries.

ISBN-13: 978–0–230–52534–4 hardback
ISBN-10: 0–230–52534–2 hardback

This book is printed on paper suitable for recycling and made from fully managed and sustained forest sources. Logging, pulping and manufacturing processes are expected to conform to the environmental regulations of the country of origin.

A catalogue record for this book is available from the British Library.

Library of Congress Cataloging-in-Publication Data

Energy efficiency and sustainable consumption: the rebound effect/
[edited by] Horace Herring and Steve Sorrell.
p. cm. – (Energy, climate and the environment)
Includes index.
ISBN 978–0–230–52534–4 (alk. paper)
1. Energy consumption. I. Herring, Horace. II. Sorrell, Steve, 1963–
HD9502.A2E543834 2004
333.79'13—dc22 2008025126

10 9 8 7 6 5 4 3 2 1
18 17 16 15 14 13 12 11 10 09

Printed and bound in Great Britain by
CPI Antony Rowe, Chippenham and Eastbourne

Contents

List of Figures

List of Tables

List of Boxes

Series Editor Preface

David Elliott

Concerns about the potential environmental, social and economic impacts of climate change have led to a major international debate over what could and should be done to reduce emissions of greenhouse gases, which are claimed to be the main cause. There is still a scientific debate over the likely scale of climate change, and the complex interactions between human activities and climate systems, but, in the words of no less than Governor Arnold Schwarzenegger, 'I say the debate is over. We know the science, we see the threat, and the time for action is now.'

Whatever we do now, there will have to be considerable social and economic adaptation to the impacts of climate change-preparing for increased flooding and other climate-related problems. However, the more fundamental response is to try to reduce or avoid the human activities that are seen as causing climate change. That means, primarily, trying to reduce or eliminate emission of greenhouse gases from the combustion of fossil fuels, for example in vehicles and power stations. Given that around 80 per cent of the energy used in the world at present comes from these sources, this will be a major technological, economic and political undertaking. It will involve reducing demand for energy (for example, via lifestyle choice changes), producing and using whatever energy we still need more efficiently (that is, getting more from less), and supplying the reduced amount of energy from non-fossil sources (basically switching over to renewables and/or nuclear power).

Each of these options opens up a range of social, economic and environmental issues. Industrial society and modern consumer cultures have been based on the ever-expanding use of relatively cheap fossil fuels, so the changes required will inevitably be challenging. Perhaps equally inevitable are disagreements and conflicts over the merits and demerits of the various options and in relation to the strategies and policies for pursuing them. These conflicts and associated debates sometimes concern technical issues, but there are usually also underlying political and ideological commitments and agendas which shape, or at least colour, the ostensibly technical debates. In particular, at times, technical assertions can be used to buttress specific policy frameworks in ways which subsequently prove to be flawed.

The aim of this series is to provide texts which lay out the technical, environmental and political issues relating to the various proposed policies for responding to climate change. The focus is not primarily on the science of climate change, or on the technological detail, although there will be accounts of the state of the art, to aid assessment of the viability of the various options. However, the main focus is the policy conflicts over which strategy to pursue. The series adopts a critical approach and attempts to identify flaws in emerging policies, propositions and assertions. In particular, it seeks to illuminate counter-intuitive assessments, conclusions and new perspectives. The aim is not simply to map the debates, but to explore their structure, their underlying assumptions and their limitations. Texts are incisive and authoritative sources of critical analysis and commentary, indicating clearly the divergent views that have emerged and also identifying the shortcomings of these views. However, the books do not simply provide overviews, or abstract reflections, they also offer concrete policy prescriptions.

The present volume looks critically at the role that the more efficient use of energy might play in mitigating climate change. While ostensibly it seems no more than common sense to believe that increasing efficiency will reduce energy use and therefore emissions, in fact this is not necessarily the outcome. Despite significant increases in energy efficiency, energy use in most sectors has continued to increase. This book looks at the historic failure to reduce consumption via improved efficiency, focusing on the importance of the so-called 'rebound effect', and at how we might do better in the future. The simple message is that achieving reductions in energy use, and consequent greenhouse gas emissions, is not just a question of technology. What also matters is how we use the technology and how we match our expectations of material consumption to the limitations of the planet's climate and ecological systems. The debate over the need for sustainable consumption is only just beginning, and as yet there are few examples of significant and successful practice. This book seeks to move the discussion on by exposing some of the fallacies about the role of energy efficiency, while at the same time addressing some of the contentious issues surrounding the concept of sustainable consumption. In particular, and controversially, it concludes that, while the promotion of energy efficiency has an important role to play in achieving a sustainable economy, it is unlikely to be sufficient while rich countries continue to pursue high levels of economic growth.

Notes on the Contributors

Grant Allan is a Research Fellow in the Fraser of Allander Institute, Department of Economics, University of Strathclyde. A graduate of Strathclyde and Edinburgh Universities, since 2004 he has undertaken research through the EPSRC SuperGen Marine programme. His research interests are in regional and national energy-economy-environment modelling.

Robert U. Ayres joined INSEAD at Fontainebleau, France in 1992 and became the first Chair of Management and the Environment, as well as the founder of CMER which he directed from 1992–2000. Now retired, he is still active as Emeritus Professor and produces numerous publications in the field of ecological and environmental economics.

Manuel Frondel is Head of the Department of Environment and Resources at RWI Essen. His research interests include applied econometrics in the fields of environmental, resource and energy economics, such as methodological issues on the evaluation of policy instruments, the analysis of energy and water demand, and the substitution of energy by non-energy inputs.

Michelle Gilmartin is currently undertaking a PhD studentship at the Fraser of Allander Institute, University of Strathclyde. A graduate of Strathclyde and Warwick Universities, her PhD studies are funded through the EPSRC SuperGen Marine programme. Her research interests are Computable General Equilibrium (CGE) modelling, macroeconomic modelling and the economics of renewable energy.

Horace Herring is a freelance writer and energy researcher, and is a Visiting Research Fellow in the Sustainable Technologies Group at the Open University, England. His research interests include energy efficiency, environmental history and sustainable consumption.

Mikko Jalas is a research fellow at the Department of Organisation and Management at Helsinki School of Economics, Finland. He received his doctoral degree in 2006 and has published a number of journal articles and book chapters on the environmental impacts of consumption and time use. His current research interests cover time politics, as well as speed and rhythm of consumption.

Roger Levett is a partner in Levett-Therivel sustainability consultants. He specialises in public policy, planning and appraisal for sustainability.

Peter G. McGregor is Professor in the Fraser of Allander Institute and Department of Economics, University of Strathclyde, as well as a member of the Centre for Public Policy for Regions, Universities of Glasgow and Strathclyde, Scotland. He has research interests in economoy-energy-environment interactions, multi-sectoral regional economic modelling and the economics of devolution.

Jørgen S. Nørgård has since 1970 worked mostly at the Technical University of Denmark, investigating and teaching how to achieve environmental sustainability through changes in technology, economics, lifestyles etc., with a focus of energy savings. Now retired, he is still an active writer and researcher.

Jörg Peters studied economics at the University of Cologne and the ENS de Cachan, Paris, and joined RWI in March 2005. His research focus is on survey and evaluation methodology and energy demand analysis. He is currently managing several research projects on the evaluation of development interventions in rural Africa.

Harry D. Saunders is Managing Director of Decision Processes Incorporated, a management consulting firm in the United States. His hobby is research. He does unfunded research in theoretical biology, criminal and constitutional law, and energy economics. His wife and four children believe this is a ridiculous hobby.

Steve Sorrell is a Senior Fellow in the Sussex Energy Group at SPRU, Sussex University. His research interests include energy efficiency in industry and commerce, climate policy and emissions trading, and the implementation of environmental regulation. He has just completed a two-year study for the UK Energy Research Centre on the 'Rebound Effect'.

J. Kim Swales is a Professor in the Department of Economics, University of Strathclyde, Glasgow, and Director of the Fraser of Allander Institute, University of Strathclyde, as well as a member of the Centre for Public Policy for Regions, Universities of Glasgow and Strathclyde.

Karen Turner is a Senior Lecturer in the Department of Economics, University of Strathclyde. She has worked with the FAI regional modelling team since 1999. Her main research area is accounting and modelling energy-economy-environment interaction using input-ouput and CGE techniques.

Colin Vance has worked since 2006 in the Department of Environment and Resources of RWI Essen. In previous positions with the US Environmental Protection Agency and the German Aerospace Centre, his research focused on spatial econometric and land use policy issues. His current work addresses the determinants of transport and energy demand in the residential sector.

Benjamin Warr joined INSEAD at Fontainebleau, France in 2001 and is now a Senior Research Fellow in the Social Innovation Centre. His interdisciplinary research involves investigation of the historical role of energy and technological progress in generating wealth.

1
Introduction

Steve Sorrell and Horace Herring

Avoiding dangerous climate change is the defining challenge for humanity in the twenty-first century. If we fail to act urgently and effectively to reduce global emissions of greenhouse gases, catastrophe looms. The Fourth Assessment Report from the Intergovernmental Panel on Climate Change (IPCC) leaves little doubt that the climate is changing; human activities are the cause and the consequences are serious (IPCC, 2007c). For example, the likely impacts of a 2–3°C increase in global mean temperatures include an additional two billion people affected by water scarcity; significant reductions in agricultural productivity and food availability in developing countries; and increased risk of extinction for 20–30 per cent of the world's species (IPCC, 2007a, 2007c). Of particular concern is the potential for triggering the irreversible disintegration of the Greenland and West Antarctica ice sheets, leading ultimately – and perhaps rapidly – to a global sea level rise of around twelve metres (Hansen, 2007).

To avoid the worst of these consequences the UN, the European Union and the UK government have endorsed the aim of keeping global average temperature increases to less than 2°C above pre-industrial levels. But to have reasonable (e.g. greater than 50 per cent) chance of meeting this goal, the concentration of greenhouse gases in the atmosphere would need to be stabilised below 450 parts per million of carbon dioxide equivalent (Meinhausen, 2005; Baer and Mastrandrea, 2006). This suggests that global emissions of greenhouse gases will need to peak sometime within the next ten to fifteen years and then be reduced by at least 50 per cent by mid-century, with further reductions beyond (by which time it is anticipated that the global population will exceed nine billion). To achieve global political agreement on such an ambitious target, developed countries will almost certainly need to commit to proportionally greater emission reductions, perhaps of the order of 80–90 per cent.

We cannot overstate the enormity of this challenge. Technically it should be achievable, but only through the rapid deployment of a wide range of low carbon energy supply technologies, in parallel with a reduction – or at least a curtailment in the growth – of global energy demand (Pacala and Socolow, 2004; IPCC, 2007b; Krewitt et al., 2007). Both independent studies and official government policies assume that the primary mechanism for reducing energy demand will be the encouragement of improved *energy efficiency* throughout the economy – for example, through encouraging the adoption of thermal insulation and low energy lighting. The technical potential for such improvements is acknowledged to be very large and in many cases the economics are extremely favourable, even without putting a price on carbon emissions. Indeed, it seems difficult to envisage a low carbon economy that was *not* characterised by much higher levels of end-use efficiency than exist today. In contrast to several energy supply options, the encouragement of cost effective energy-efficiency measures also receives relatively uncritical support from business, environmental groups, political parties and the general public. In modern economies 'efficiency has become nearly synonymous with "good" . . . so internalised that one hardly thinks about it, let alone questions it' (Princen, 2005).

But there is a flaw in this efficiency argument. Improving energy efficiency may not be as effective in reducing energy demand as is generally assumed. Something called 'the rebound effect'– or more precisely 'rebound effects' – may get in the way and reduce the size of the 'energy savings' achieved. Some heretics even argue that energy-efficiency improvements will lead to *increased* energy demand, at least over the long term. Encouraging energy efficiency as a means of reducing carbon emissions would then be futile – rather like a dog chasing its tail. In the past such arguments were usually dismissed on the grounds that rebound effects have been investigated and found to be largely irrelevant (Lovins et al., 1988). But is this really the case? Do we really have a clear understanding of what these effects are, how they operate and how important they are in different circumstances? Are we overestimating the 'energy savings' that improved energy efficiency can provide? Could the heretics be right – at least in some cases? If so, what are those cases and how can we identify them? Are such cases the norm? Most importantly, why should something as potentially important as rebound effects continue to be so widely overlooked?

These are the questions that this book sets out to address. It seeks to present the latest research on rebound effects, while at the same time making the issues as accessible as possible to a non-technical audience. The majority of the chapters draw upon concepts and techniques from economics, but with a focus that ranges from individual behaviour to the determinants of long-run economic growth. The book also includes four contributions that discuss the broader issues associated with energy efficiency and sustainability, including in particular the complementary notion of 'sufficiency' (Princen, 2005). A recurring theme is how partial and limited our present understanding of rebound effects is. Hence a primary objective of the book is to raise awareness and improve understanding of this topic and to encourage much wider debate and research.

We expect that many readers will be relatively unfamiliar with the notion of rebound effects. Moreover, experience suggests that many of those who think they are familiar may in fact be confused over key issues. The following two sections therefore classify the different types of rebound effect, provide a brief historical context and elaborate on the definition and measurement of energy efficiency and energy consumption. The final section provides an overview of the book and a brief summary of each of the chapters.

What are rebound effects?

To achieve reductions in carbon emissions, most governments are seeking ways to improve energy efficiency throughout the economy. It is generally assumed that such improvements will reduce overall energy consumption, at least compared to a scenario in which such improvements are not made. But for many years, economists have recognised that a range of mechanisms, commonly grouped under the heading of *rebound effects*, may reduce the size of the 'energy savings' achieved. Indeed, some economists have argued that the introduction of certain types of energy-efficient technology in the past has contributed to an overall increase in energy demand – an outcome that has been termed 'backfire'. This may apply in particular to pervasive new technologies, such as steam engines in the nineteenth century, that significantly raise overall economic productivity as well as improving energy efficiency (Alcott, 2005).

If these rebound effects are significant, they could have far-reaching implications for energy and climate policy. While cost-effective improvements in energy efficiency should improve welfare and increase economic output, they could in some cases provide an ineffective or even a counterproductive means of tackling climate change. However, even if backfire is possible or indeed common, it does not necessarily follow that *all* improvements in energy efficiency will increase overall energy consumption or in particular that the improvements encouraged by policy measures will do so.

The nature, operation and importance of rebound effects are the focus of a long-running debate within energy economics. On the micro level, the question is whether improvements in the technical efficiency of energy use can be expected to reduce energy consumption by the amount predicted by simple engineering calculations. For example, will a 20 per cent improvement in the fuel efficiency of passenger cars lead to a corresponding 20 per cent reduction in motor-fuel consumption for personal automotive travel? Simple economic theory suggests that it will not. Since energy-efficiency improvements reduce the marginal cost of energy services such as travel, the consumption of those services may be expected to increase. For example, since the cost per kilometre of driving is cheaper, consumers may choose to drive further and/or more often. This increased consumption of energy services may be expected to offset some of the predicted reduction in energy consumption.

This so-called *direct rebound effect* was first brought to the attention of energy economists by Daniel Khazzoom (1980) and has since been the focus of much research (Greening et al., 2000). But even if there is no direct rebound effect for a particular energy service (e.g. even if consumers choose not to drive any further in their fuel-efficient car), there are a number of other reasons why the economy-wide reduction in energy consumption may be less than simple calculations suggest. For example, the money saved on motor-fuel consumption may be spent on other goods and services that also require energy to provide – such as an overseas holiday. These so-called *indirect rebound effects* can take a number of forms that are briefly outlined in Box 1.1. Both direct and indirect rebound effects apply equally to energy-efficiency improvements by consumers, such as the purchase of a more fuel-efficient car (Figure 1.1), and energy-efficiency improvements by producers, such as the adoption of energy-efficient process technology (Figure 1.2).

The *overall* or *economy-wide* rebound effect from an energy-efficiency improvement represents the sum of these direct and indirect effects. It is normally expressed as a percentage of the *expected* energy savings from an energy-efficiency improvement. Hence, an economy-wide rebound effect of 20 per cent means that 20 per cent of the potential energy

Box 1.1 Indirect rebound effects

- The equipment used to improve energy efficiency (e.g. thermal insulation) will itself require energy to manufacture and install and this 'embodied' energy consumption will offset some of the energy savings achieved.
- Consumers may use the cost savings from energy-efficiency improvements to purchase other goods and services which themselves require energy to provide. For example, the cost savings from a more energy-efficient central heating system may be put towards an overseas holiday.
- Producers may use the cost savings from energy-efficiency improvements to increase output, thereby increasing consumption of capital, labour and materials which themselves require energy to provide. If the energy-efficiency improvements are sector-wide, they may lead to lower product prices, increased consumption of the relevant products and further increases in energy consumption.
- Cost-effective energy-efficiency improvements will increase the overall productivity of the economy, thereby encouraging economic growth. The increased consumption of goods and services may in turn drive up energy consumption.
- Large-scale reductions in energy demand may translate into lower energy prices which will encourage energy consumption to increase. The reduction in energy prices will also increase real income, thereby encouraging investment and generating an extra stimulus to aggregate output and energy use.
- Both the energy-efficiency improvements and the associated reductions in energy prices will reduce the cost of energy-intensive goods and services to a greater extent than non-energy-intensive goods and services, thereby encouraging consumer demand to shift towards the former.

savings are 'taken back' through one or more of the mechanisms indicated above. A rebound effect of 100 per cent means that the expected energy savings are entirely offset, leading to zero net savings. Backfire means that rebound effects exceed 100 per cent, leading to increased energy consumption.

Rebound effects need to be defined in relation to a particular *time frame* (e.g. short, medium or long term) and *system boundary* for the

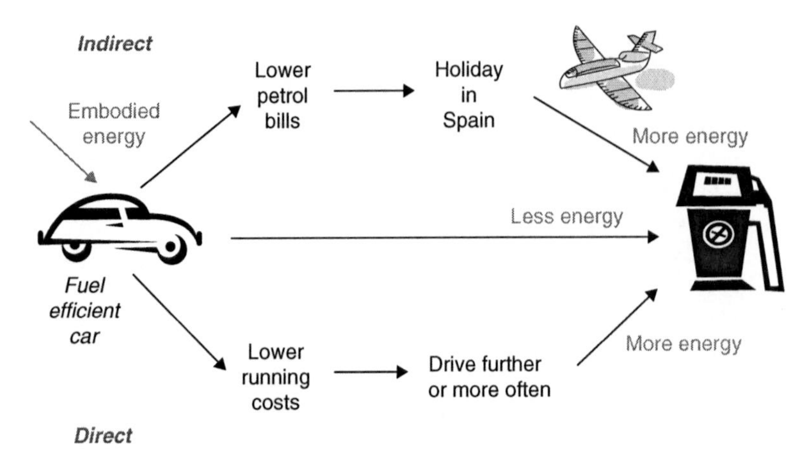

Figure 1.1 Illustration of rebound effects for consumers

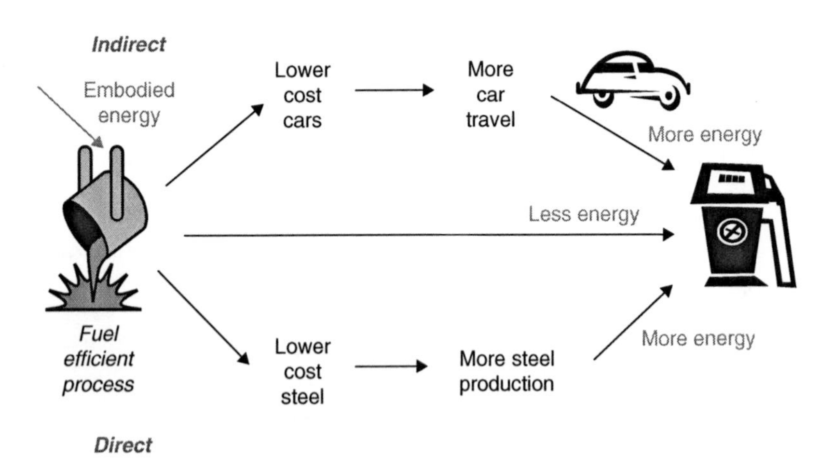

Figure 1.2 Illustration of rebound effects for producers

relevant energy consumption (e.g. household, firm, sector, national economy). For example, energy savings may be expected to be smaller for the economy as a whole than for the individual household or firm that is implementing an energy-efficiency improvement. The economy-wide effect is normally defined in relation to a national economy, but if energy-efficiency improvements lead to changes in trade patterns and international energy prices there may also be effects in other countries. Rebound effects may also be expected to increase in importance over time as markets, technology and behaviour adjust. From a climate change

Box 1.2 Classifying rebound effects

The economy-wide rebound effect represents the sum of the direct and indirect effects. For energy-efficiency improvements by consumers, it is helpful to decompose the direct rebound effect into:

(a) *a substitution effect,* whereby consumption of the (cheaper) energy service substitutes for the consumption of other goods and services while maintaining a constant level of 'utility', or consumer satisfaction; and
(b) an *income effect,* whereby the increase in real income achieved by the energy-efficiency improvement allows a higher level of utility to be achieved by increasing consumption of all goods and services, including the energy service.

Similarly, the direct rebound effect for producers may be decomposed into:

(a) a *substitution effect,* whereby the cheaper energy service substitutes for the use of capital, labour and materials in producing a constant level of output; and
(b) an *output effect,* whereby the cost savings from the energy-efficiency improvement allows a higher level of output to be produced – thereby increasing consumption of all inputs, including the energy service.

It is also helpful to decompose the indirect rebound effect into:

(a) the *embodied energy,* or indirect energy consumption required to *achieve* the energy-efficiency improvement, such as the energy required to produce and install thermal insulation; and
(b) the *secondary effects* that result as a *consequence* of the energy-efficiency improvement, which include the mechanisms listed in Box 1.1.

The relative size of each effect may vary widely from one circumstance to another and in some cases individual components of the rebound effect may be negative.

perspective, it is the long-term effect on global energy consumption that is most relevant, but this is also the effect that is hardest to estimate.

Box 1.2 provides a general classification scheme for rebound effects, which is illustrated diagrammatically in Figure 1.3. The economy-wide rebound effect represents the net effect of a number of different

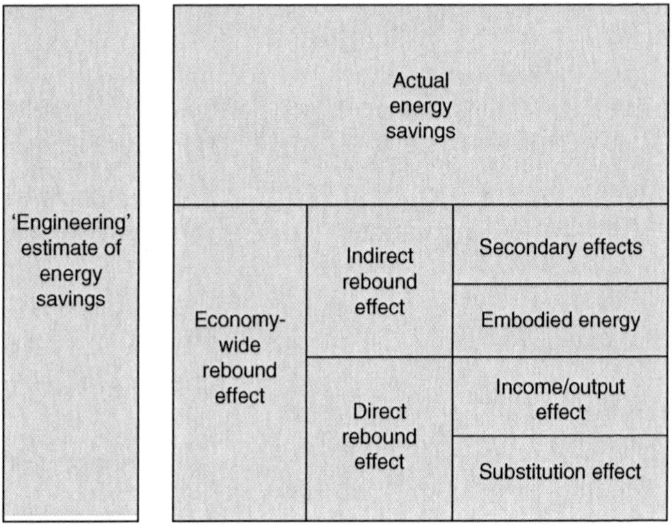

Figure 1.3 Classification scheme for rebound effects

mechanisms that are individually complex, mutually interdependent and likely to vary in importance from one type of energy-efficiency improvement to another. Estimating the magnitude of the economy-wide rebound effect in any particular instance is therefore very difficult. Nevertheless, the general implication is that the energy 'saved' from energy-efficiency improvements will be less than is conventionally assumed.

The view that economically justified energy-efficiency improvements will increase rather than reduce energy consumption was first put forward by the British economist, William Stanley Jevons, as long ago as 1865. It has subsequently become known as 'Jevons' Paradox' and also as the 'Khazzoom-Brookes (K-B) postulate', after two contemporary economists (Len Brookes and Daniel Khazzoom) who have been closely associated with this idea. In its original formulation, the K-B postulate states that 'with fixed real energy prices, energy-efficiency gains will *increase* energy consumption above what it would be without those gains' (Saunders, 1992).

If it were correct, the K-B postulate would have deeply troubling policy implications. It would imply that many of the policies used to promote energy efficiency may reduce neither energy consumption nor carbon emissions. The conventional assumptions of energy analysts, policy-makers, business and lay people alike would be turned on their head. Alternatively, even if the postulate were incorrect, the various mechanisms described above could still make energy-efficiency policies less effective in reducing energy consumption than is commonly assumed. In either case, the rebound effect could have important implications for global efforts to address climate change. But despite its potential importance, the topic is widely neglected and shrouded in controversy and confusion.

Rebound effects tend to be almost universally ignored in official analyses of the potential energy savings from energy-efficiency improvements. A rare exception is UK policy to improve the thermal insulation of housing, where it is expected that some of the benefits will be taken as higher internal temperatures rather than reduced energy consumption (DEFRA, 2007). But the direct rebound effects for other energy-efficiency measures are generally ignored, as are the potential indirect effects for all measures. Much the same applies to energy modelling studies and to independent estimates of energy-efficiency potentials. For example, the Stern Review of the economics of climate change overlooks rebound effects altogether (Stern, 2007), while the Fourth Assessment Report from the Intergovernmental Panel on Climate Change simply notes that the literature is divided on the magnitude of the effect (IPCC, 2007b).

While energy analysts recognise that direct and indirect rebound effects may reduce the energy savings from energy-efficiency improvements, there is dispute over how important these effects are – both individually and in combination. Some argue that rebound effects are of minor importance for most energy services, largely because the demand for many of these services appears to be relatively unresponsive to changes in prices and because energy typically forms a small share of the total costs (Lovins et al., 1988; Lovins, 1998; Schipper and Grubb, 2000). Others argue that they are sufficiently important to completely offset the energy savings from improved energy efficiency (Brookes, 2000). At first sight, it appears odd for competent and experienced analysts to hold such widely diverging views on what appears to be an empirical question. This suggests one of three things: first, different authors may be using different definitions of rebound effects, together with different definitions of associated issues such as the relevant system boundaries; second, the empirical evidence for rebound effects may be sufficiently sparse, ambiguous and inconclusive to be open to widely varying interpretations; or third, fundamental assumptions regarding how the economy operates may be in dispute.

In practice, all of these factors appear to be relevant. The empirical evidence for direct rebound effects is very patchy, mainly focused on a limited number of consumer energy services such as personal automotive transport and almost wholly confined to OECD economies. Quantitative estimates of indirect and economy-wide rebound effects are difficult to make and consequently rare, leaving authors such as Brookes (2000) to rely on a mix of theoretical argument, illustrative examples and 'suggestive' evidence from econometric analysis and economic history. Much of this evidence rests upon theoretical assumptions and methodological approaches that are both highly technical and openly contested. There is also confusion, misunderstanding and disagreement over basic definitional issues such as the meaning of 'improvements in energy efficiency'. The result is that commentators talk past one another, with disagreements originating in part because different people are talking about different things. For example, some commentators appear to equate 'the' rebound effect solely with the direct effect, and thereby ignore the indirect effects that are the primary concern of economists such as Jevons. In view of this, it is helpful to look at some of these definitional issues more closely.

Energy efficiency and energy savings

Energy-efficiency improvements are generally assumed to reduce energy consumption below where it would have been without those improvements. Rebound effects reduce the size of these energy savings. However, estimating the size of any 'energy savings' is far from straightforward, since:

- Real-world economies do not permit controlled experiments, so the relationship between a change in energy efficiency and a subsequent change in energy consumption is likely to be mediated by a host of confounding variables.
- We can't observe what energy consumption 'would have been' without the energy-efficiency improvement, so the estimated 'savings' will always be uncertain.
- Energy efficiency is not controlled externally by an experimenter and may be influenced by a variety of technical, economic and policy variables. In particular, the direction of causality may run in reverse – with changes in energy consumption (whatever their cause) leading to changes in different measures of energy efficiency.

Energy efficiency may be defined as the ratio of useful outputs to energy inputs for a system. The system in question may be an individual energy conversion device (e.g. a boiler), a building, an industrial process, a firm, a sector or an entire economy. In all cases, the measure of energy efficiency will depend upon how 'useful' is defined and how inputs and outputs are measured (Patterson, 1996). The options include:

- *Thermodynamic measures*: where the outputs are defined in terms of either heat content or the capacity to perform useful work;
- *Physical measures*: where the outputs are defined in physical terms, such as vehicle kilometres or tonnes of steel; or
- *Economic measures*: where the outputs (and sometimes also the inputs) are defined in economic terms, such as value-added or GDP.

When outputs are measured in thermodynamic or physical terms, the term 'energy efficiency' tends to be used, but when outputs are measured in economic terms it is more common to use the term 'energy productivity'. The inverse of both measures is termed 'energy intensity'. The choice of measures for inputs and outputs, the appropriate system boundaries and the time frame under consideration can vary widely from one study to another. However, physical and economic measures of energy efficiency tend to be influenced by a greater range of variables than thermodynamic measures, as do measures appropriate to wider system boundaries. Hence, the indicator that is furthest from a thermodynamic measure of energy efficiency is the ratio of GDP to total primary energy consumption within a national economy. The conclusions drawn regarding the magnitude and importance of rebound effects will therefore depend upon the particular measure of energy efficiency that is chosen.

Economists are primarily interested in energy-efficiency improvements that are consistent with the best use of all economic resources. These are conventionally divided into two categories: those that are associated with improvements in overall, or 'total factor' productivity ('technical change'), and those that are not ('substitution'). The consequences of technical change are of particular interest, since this contributes to the growth in economic output. However, distinguishing empirically between these two categories can be challenging.

Increases in energy prices may lead to improvements in the energy efficiency of a system through the *substitution* of capital or labour inputs for energy. For example, thermal insulation may reduce the consumption of gas for space heating. But if the prices of other inputs are unchanged, the total cost of producing a given level of output will have increased. In contrast, improvements in both energy productivity and overall or total factor productivity may result from *technical change*. This is conventionally assumed to occur independently of any change in relative prices and is considered desirable since it occurs without any reduction in economic output.

The importance of these definitions for estimates of the rebound effect is not often recognised. Many commentators assume that the relevant independent variable for the rebound effect is an improvement in the thermodynamic efficiency of individual conversion devices or industrial processes. But such improvements will only translate into comparable improvements in different measures of energy efficiency, or measures of energy efficiency applicable to wider system boundaries, if several of the mechanisms responsible for the rebound effect fail to come into play. For example, improvements in the number of litres used per vehicle kilometre will only translate into improvements in the number of litres used per passenger kilometre if there are no associated changes in average vehicle load factors.

Rebound effects may be expected to increase over time and with the widening of the system boundary for the dependent variable (energy consumption). Hence, to capture the full range of rebound effects, the system boundary for the independent variable (energy efficiency) should be relatively narrow, while the system boundary for the dependent variable should be as wide as possible. For example, the independent variable could be the energy efficiency of an electric motor, while the dependent variable could be economy-wide energy consumption. However, measuring or estimating the economy-wide effects of such micro-level changes effects is, at best, challenging. Partly for this reason, the independent variable for many studies of rebound effects is a measure of energy efficiency that is applicable to relatively wide system boundaries – such as an industrial sector. But such studies may overlook 'lower-level' rebound effects. For example, improvements in the energy efficiency of electric motors in the engineering sector may lead to rebound effects within that sector, with the result that the energy intensity of that sector is reduced by less than it would be in the absence of such effects. But if the energy intensity of the sector is taken as the independent variable, these lower-level rebound effects will be missed.

Aggregate measures of energy efficiency will also depend upon how different types of energy input are combined. While it is common practice to aggregate different energy types on the basis of their heat content, this neglects the 'thermodynamic quality' of each energy type, or its ability to perform useful work. The latter, in turn, is only one of several factors that determine the economic productivity of different energy types, with others including cleanliness, amenability to storage, safety, flexibility and the use to which the energy is put (Cleveland et al., 2000). These differences are broadly reflected in the relative prices of different energy types and these prices can be used to create a 'quality adjusted' measure of aggregate energy consumption (Berndt, 1978; Kaufmann, 1994). In general, when the 'quality' of energy inputs is accounted for, aggregate measures of energy efficiency are found to be improving more slowly than is commonly supposed (Hong, 1983; Zarnikau, 1999; Cleveland et al., 2000).

Improvements in any measure of energy efficiency rarely occur in isolation but are typically associated with broader improvements in the productivity of other inputs. This may be the case even when the primary objective of the relevant investment or behavioural change is to reduce energy costs (Pye and McKane, 1998). Importantly, if the total cost savings exceed the saving in energy costs, then any rebound effects may be amplified.

In summary, disagreement over the nature and size of rebound effects may result in part from different definitions of the relevant independent variable (energy efficiency) and in part from different choices for the time frame and system boundary for the dependent variable (energy consumption). While there is no single 'right' choice, the selection of particular measures could lead to some types of rebound effect being either underestimated or overlooked.

Overview of the book

The book is divided into three parts: 1. Direct rebound effects; 2. Economy-wide rebound effects; 3. Rebound effects and sustainable consumption. The contents of each are briefly summarised below.

Part I: Direct rebound effects

Part I examines direct rebound effects. As well as being the best-known and most easily understood component of the economy-wide rebound effect, this is also the area that has received the most research. Direct rebound effects relate to individual energy services, such as heating, lighting and refrigeration and are confined to the energy required to provide that service. The primary issue here is the extent to which demand for those services may increase following an improvement in energy efficiency. While this is a generic issue for all energy services, the research to date has been overwhelmingly confined to personal transport and household heating in the United States.

Chapter 2 by Steve Sorrell provides an overview of the methodological approaches to estimating direct rebound effects and reviews the evidence that is currently available. Sorrell highlights the diversity of the existing evidence base, the methodological difficulties associated with such estimates, the potential sources of bias and the possibility that several studies have overestimated the effect. For personal automotive transport, household heating and household cooling in OECD countries, the evidence suggests a long-run direct rebound effect of 30 per cent or less. The direct rebound effect for other household energy services is expected to be smaller. However, these conclusions are subject to a number of qualifications and provide no guidance on the magnitude of rebound effects for producers or for households in developing countries.

Chapter 3 by Manuel Frondel, Jörg Peters and Colin Vance describes an econometric estimate of the direct rebound effect for passenger transport in Germany. The chapter introduces three different measures of the direct rebound effect and estimates each of these using travel diary data collected from a panel of German households between 1997 and 2005. Frondel and his colleagues estimate the long-run direct rebound effect to be between 57 and 67 per cent, which is higher than the consensus in the literature. This leads them to question the effectiveness of fuel efficiency standards as a means to reduce transport fuel consumption.

Part II: Economy-wide rebound effects

Part II includes four chapters that examine the evidence for economy-wide rebound effects as well as the relationship between these effects and the determinants of long-run economic growth. Economy-wide rebound effects are necessarily more complex than direct effects and are very difficult to estimate empirically. These chapters therefore place more reliance upon theoretical arguments and energy-economic modelling, as well as highlighting various 'indirect' sources of evidence that could provide some indication of the magnitude of economy-wide rebound effects. Two of the chapters argue that energy-efficiency improvements may play a central role in driving economic growth.

Chapter 4 by Grant Allan and colleagues describes the use of a computable general equilibrium (CGE) model of the UK economy to estimate the impact of a 5 per cent across the board improvement in the energy efficiency of all production sectors, including energy supply. In their baseline scenario, they estimate the long-run, economy-wide rebound effect to be 14 per cent for electricity and 31 per cent for fuel. However, these results are found to be sensitive to a number of variables, including the assumed scope for substitution between energy and other inputs.

Chapter 5 by Harry Saunders describes how rebound effects may be explored with the help of neoclassical production theory. Saunders provides a non-technical account of the use of this theory to describe producer behaviour, the representation of energy-efficiency improvements in 'production functions' and the use of those functions to simulate the impact of particular technologies. He demonstrates how a simple and widely used production function leads to the prediction that energy-efficiency improvements will increase energy consumption.

Chapter 6 by Robert Ayres and Benjamin Warr examines economy-wide rebound effects from the perspective of ecological economics. Ayres and Warr present two theses that run directly counter to conventional wisdom. The first is that increasing energy efficiency has been a major – perhaps the major – driver of economic growth since the industrial revolution and that 'technological progress' as normally construed by economists is primarily due to increasing thermodynamic efficiency. The second is that, while reduced carbon dioxide emissions are essential for long-term global sustainability, increasing the cost of energy by introducing a carbon tax could have an adverse impact on economic growth. These propositions are supported with reference to an unconventional but empirically validated model of economic growth in the United States.

In Chapter 7, Steve Sorrell provides a summary and critique of the arguments and evidence used in support of Jevons' Paradox. He provides a historical context to the debate, summarises the arguments in favour of backfire provided by Len Brookes, identifies some empirical and theoretical weaknesses with these arguments and examines whether more recent research supports Brookes' claims. He then summarises the lessons to be learned from neoclassical production and growth theory, highlighting the dependence of these results on specific theoretical assumptions and the questions it raises for standard economic methodologies. Sorrell also highlights some interesting and important parallels between Brookes' arguments and those of contemporary ecological economists. A core theme is that energy – and by implication improved energy efficiency – may play a more important role in driving economic growth than is conventionally assumed.

Part III: Rebound effects and sustainable consumption

Part III moves beyond economic concepts and methodologies to examine the social, political and cultural issues relevant to energy efficiency and sustainability. A core theme is the idea of 'sufficiency', defined broadly as limiting the consumption of goods and services either to better satisfy personal desires such as health or meaningful work or to contribute to collective goals such as long-term environmental sustainability. While improved energy efficiency reduces energy consumption per unit of energy service, sufficiency focuses on limiting consumption of the services themselves. The concept of sufficiency resonates strongly with long-standing critiques of the sustainability of economic growth and could potentially offer a means to mitigate the negative environmental impacts of rebound effects. However, current policy strategies to encourage sufficiency focus largely on encouraging voluntary action and remain both limited and ineffective.

In Chapter 8, Mikko Jalas argues that a focus on the *time* associated with different household activities can change the terms of the rebound debate, raise new questions and highlight some potentially important but hitherto neglected mechanisms. As an illustration, Jalas compares the outsourcing of household activities (e.g. lawn mowing) to the conventional ownership and use of the relevant goods (e.g. lawn mowers). This type of organisational innovation is frequently claimed to improve energy efficiency. However, Jalas shows how such 'time-saving' arrangements could also encourage increases in the hours of paid work and hence economic output, which would run counter to the objectives of sufficiency.

In Chapter 9, Roger Levett criticises the conventional assumption that there is some sensible, predictable relationship between a policy intervention and its results – as exemplified by the notion of 'evidence-based policy'. The energy-efficiency rebound effect is simply one example of the unintended outcomes and vicious circles that can result from interventions in complex economic and social systems. These tend to be characterised by complexity, unpredictability, irreversibility, path dependence and various types of positive and negative feedback mechanisms. He outlines a 'systems literate' approach to policy-making that includes principles for managing rebounds and other types of unwanted feedback effect.

Chapter 10, by Jørgen Nørgård places the debate over the rebound effect in the context of a broader critique of dominant economic assumptions. He argues that efforts to reduce environmental damage through improved energy efficiency will continue to be more than offset by the negative effects of continued economic growth – itself partly originating from the productivity improvements associated with improved energy efficiency. He notes that sustainability in developed countries is likely to require a reduction in per capita energy and resource consumption of as much as 90 per cent. While improved energy efficiency can play a significant role in achieving this, it needs to be combined with an ethic of sufficiency in the use of energy services. Nørgård argues that such changes will also be beneficial to human well-being.

In Chapter 11, Horace Herring reviews the strengths and weaknesses of the 'sufficiency strategy' as a means to mitigate the negative environmental consequences of rebound effects. While endorsing the broad principle, he highlights the limitations of voluntary action to reduce consumption, due in part to the existence of rebound effects. On one level, this points to the need to focus primarily on desired ends (absolute reductions in greenhouse gas emissions) rather than specific means (efficiency or sufficiency) as exemplified by international agreement on emissions targets. On a more fundamental level, this highlights the contradictions between continued economic growth and environmental sustainability. Herring argues that sustainability cannot be achieved by a minority of ethical consumers acting alone: it needs the actions of the majority, expressed through government policy.

Conclusions

Finally, in Chapter 12 Sorrell and Herring summarise the main arguments of the book and highlight some of the implications for energy and climate policy. They do so by advancing six propositions, namely:

- rebound effects matter;
- rebound effects can be quantified;
- policy can mitigate rebound effects;
- Jevons' Paradox holds in important cases;
- efficiency needs to be combined with sufficiency; and
- sustainability is incompatible with continued economic growth in developed countries.

The first of these is the least controversial, least normative and best supported by the available evidence, while the opposite is the case for the last. The degree of support for these propositions may therefore be expected to decline as we go down the list. However, we argue that the existence of non-trivial rebound effects makes it all the more important that such fundamental questions about the sustainability of economic growth be asked.

Acknowledgements

Much of this chapter is based upon a comprehensive review of the evidence for rebound effects, conducted by the UK Energy Research Centre and summarised in Sorrell (2007). The financial support of the UK Research Councils is gratefully acknowledged.

References

Alcott, B. (2005) 'Jevons' paradox', *Ecological Economics*, 54(1): 9–21.

Baer, P. and M. Mastrandrea (2006) *High Stakes: Designing Emissions Pathways to Reduce the Risk of Dangerous Climate Change*, London: Institute for Public Policy Research.

Berndt, E. R. (1978) 'Aggregate energy, efficiency and productivity measurement', *Annual Review of Energy*, 3: 225–73.

Brookes, L. G. (2000) 'Energy efficiency fallacies revisited', *Energy Policy*, 28(6–7): 355–66.

Cleveland, C. J., R. K. Kaufmann and D. I. Stern (2000) 'Aggregation and the role of energy in the economy', *Ecological Economics*, 32: 301–17.

DEFRA, (2007) 'Consultation document: energy, cost and carbon savings for the draft EEC 2008–11 illustrative mix', London: Department of Environment, Food and Rural Affairs.

Greening, L. A., D. L. Greene and C. Difiglio (2000) 'Energy efficiency and consumption – the rebound effect – a survey', *Energy Policy*, 28(6–7): 389–401.

Hansen, J. (2007) 'Scientific reticence and sea level rise', *Environmental Research Letters*, 2: 1–6.

Hong, N. V. (1983) 'Two measures of aggregate energy production elasticities', *Energy Journal*, 4(2): 172–7.

IPCC (2007a) 'Climate Change 2007: Impacts, Adaptation and Vulnerability', Working Group II contribution to the IPCC Fourth Assessment Report, Intergovernmental Panel on Climate Change.

IPCC (2007b) 'Climate Change 2007: Mitigation of Climate Change', Working Group III contribution to the IPCC Fourth Assessment Report, Intergovernmental Panel on Climate Change.

IPCC (2007c) 'Climate Change 2007: Synthesis Report', Fourth Assessment Report, Intergovernmental Panel on Climate Change.

Kaufmann, R. K. (1994) 'The relation between marginal product and price in US energy markets: implications for climate change policy', *Energy Economics*, 16(2): 145–58.

Khazzoom, J. D. (1980) 'Economic implications of mandated efficiency in standards for household appliances', *Energy Journal*, 1(4): 21–40.

Krewitt, W., S. Simon, W. Graus, S. Teske, A. Zervos and O. Schäfer (2007) 'The 2°C scenario – a sustainable world energy perspective', *Energy Policy*, 35(10): 4969–80.

Lovins, A. B. (1998) 'Further comments on red herrings', Letter to the *New Scientist*, No. 2152, 18 September.

Lovins, A. B., J. Henly, H. Ruderman and M. D. Levine (1988) 'Energy saving resulting from the adoption of more efficient appliances: another view; a follow-up', *Energy Journal*, 9(2): 155.

Meinhausen, M. (2005) 'What does a 2°C target mean for greenhouse gas concentrations? A brief analysis based on multi emission pathways and several climate sensitivity uncertainty estimates', in H. Schellnhuber, W. Cramer, N. Nakicenovik, T. Wigley and G. Yohe (eds), *Avoiding Dangerous Climate Change*, Cambridge: Cambridge University Press.
Pacala, S. and R. Socolow (2004) 'Stabilisation wedges: solving the climate problem for the next 50 years with current technologies', *Science*, 305: 968–72.
Patterson, M. G. (1996) 'What is energy efficiency? Concepts, indicators and methodological issues', *Energy Policy*, 24(5): 377–90.
Princen, T. (2005) *The Logic of Sufficiency*, Boston: MIT Press.
Pye, M. and A. McKane (1998) *Enhancing Shareholder Value: Making a More Compelling Energy Efficiency Case to Industry by Quantifying Non-energy Benefits*, Proceedings 1999 Summer Study on Energy Efficiency in Industry, Washington, DC.
Saunders, H. D. (1992) 'The Khazzoom-Brookes postulate and neoclassical growth', *Energy Journal*, 13(4): 131.
Schipper, L. and M. Grubb (2000) 'On the rebound? Feedback between energy intensities and energy uses in IEA countries', *Energy Policy*, 28(6–7): 367–88.
Sorrell, S. (2007) *The Rebound Effect: an Assessment of the Evidence for Economy-wide Energy Savings from Improved Energy Efficiency*, London: UK Energy Research Centre.
Stern, N. (2007) 'Stern Review: the Economics of Climate Change', London: HM Treasury.
Zarnikau, J. (1999) 'Will tomorrow's energy efficiency indices prove useful in economic studies?' *Energy Journal*, 20(3): 139–45.

Part I
Direct Rebound Effects

2
The Evidence for Direct Rebound Effects

Steve Sorrell

Direct rebound effects relate to individual energy services, such as heating, lighting and refrigeration and are confined to the energy required to provide that service. Since improved energy efficiency will reduce the marginal cost of supplying the relevant service it could lead to an increase in the consumption of that service. For example, consumers may choose to drive further and/or more often following the purchase of a fuel-efficient car because the operating cost per kilometre has fallen. Similarly, consumers may choose to heat their homes for longer periods and/or to a higher temperature following the installation of loft insulation, because the operating cost per square metre has fallen. The extent to which this occurs may be expected to vary widely from one energy service to another, from one circumstance to another and from one time period to another. But any increase in energy service consumption will reduce the 'energy savings' achieved by the energy-efficiency improvement. In some circumstances it could offset those savings altogether (backfire).

Direct rebound effects are the most familiar and widely studied component of the overall or economy-wide rebound effect. Beginning with Khazzoom (1980), there have been a series of studies estimating the size of the direct rebound effect for different energy services (Greening and Greene, 1998). These studies are extremely diverse in terms of the definitions, methodological approaches and data sources used. Also, despite growing research activity, the evidence remains sparse, inconsistent and overwhelmingly confined to a limited number of consumer energy services in the United States – notably personal automotive transport and household heating. The main reason for this is the lack of suitable data sources for other types of energy service in other sectors and countries. In addition, interpretation of the evidence is greatly hampered by the use of competing definitions, measures, terminology and notation. Many of the relevant studies do not mention the direct rebound effect at all, but nevertheless provide elasticity estimates that may, under certain assumptions, be used as proxy measures of that effect. Taken together, these features inhibit understanding of the direct rebound effect and the appropriate methodological approach to estimating its magnitude in different circumstances, and make it difficult to identify the relevance of particular studies.

This chapter provides an overview of the methodological approaches to estimating direct rebound effects and reviews the evidence that is currently available. The chapter focuses entirely on energy services in the household sector, since this is where practically all of the research has been undertaken. As a result, the conclusions do not provide guidance on either the magnitude of direct rebound effects in other sectors (where the potential for large effects may be greater) or the magnitude of the economy-wide rebound effect.

Section 1 describes the operation of the direct rebound effect, highlighting some key issues concerning the measurement of this effect and the conditions under which it may be expected to be larger or smaller. Sections 2 and 3 describe the 'quasi-experimental' and 'econometric' approaches to estimating direct rebound effects and the methodological challenges associated with each. Section 4 summarises the results of a number of studies that use this approach to estimate direct rebound effects for personal automotive transport, household heating and a limited number of other consumer energy services. Section 5 discusses a number of potential sources of bias with econometric estimates that may lead the direct rebound effect to be overestimated. Section 6 concludes.

Understanding the direct rebound effect

Energy services such as heating and lighting are provided through energy systems that involve particular combinations of capital equipment, labour, materials and marketable energy commodities such as electricity. The relevant systems may include primary conversion equipment such as boilers, secondary conversion equipment such as radiators, equipment for distributing energy and manual or electronic controls. For space heating and lighting the relevant energy systems may also include building fabric, thermal insulation, ventilation systems and glazing. The energy efficiency of such systems (ε) is defined as the ratio of useful energy outputs (S) to energy inputs (E) and may be influenced by a variety of factors other than the thermodynamic efficiency of particular conversion equipment. Different measures of energy efficiency can be developed for different system boundaries (e.g. the boiler or the house) and the magnitude of such indicators will depend upon how the energy inputs and outputs are defined and measured.

All energy services involve the transfer of either heat or work. Here, we use the term *useful work* in a generic sense to refer to output energy that serves a useful purpose. This may be measured by a variety of thermodynamic or physical indicators, which will vary from one system to another (Patterson, 1996). For example, the useful work delivered by passenger cars may be measured in terms of vehicle kilometres, passenger kilometres or (rather unconventionally) tonne kilometres.

Energy services may also have broader attributes that may be combined with useful work in a variety of ways. For example, all cars deliver passenger kilometres, but they may vary widely in terms of speed, comfort, acceleration and prestige. Consumers and producers may therefore make trade-offs between useful work and other attributes of an energy service; between energy, capital and other market goods in the production of an of energy service; and between different types of energy service.

The energy cost of useful work (P_S) may be defined as: $P_S = P_E/\varepsilon$, where P_E is the price of energy inputs. This forms one component of the generalised cost of useful work (P_G), which also includes annualised capital, maintenance and time costs. Energy efficiency improvements reduce the energy cost and hence the generalised cost of useful work. Over time, this may lead to an increase in the number of energy conversion devices, their average size, their average utilisation and/or their average load factor. For example, people may buy more cars, buy larger cars, drive them further and/or share them less. Similarly, people may buy more washing machines, buy larger machines, use them more frequently and/or increase the size of the average load. The relative importance of these variables may be expected to vary widely between different energy services and over time. For example, technological improvements in the energy efficiency of new refrigerators are unlikely to increase the average utilisation of the refrigerator stock (measured in hours/year) but could lead to a long-term increase in both the number of refrigerators and their average size (since the cost per cubic metre of refrigeration has fallen). Over the very long term, the lower cost of energy services may contribute to fundamental changes in technologies, infrastructures and lifestyles – such as a shift towards car-based commuting and increasing distances between residential, workplace and retail locations. But as the time horizon extends, the effect of such changes on the demand for the energy service becomes increasingly difficult to separate from the effect of income growth and other factors.

The estimated size of the direct rebound effect will depend upon how useful work and hence energy efficiency is defined. For example, the majority of estimates of the direct rebound effect for personal automotive transport measure useful work in terms of vehicle kilometres travelled, which is sometimes decomposed into the product of the number of vehicles and the mean distance travelled per vehicle per year (Greene et al., 1999; Small and Van Dender, 2005). Energy efficiency is then defined as vehicle kilometres per litre of fuel and rebound effects are measured as increases in distance driven. But this overlooks any changes in mean vehicle size and weight that result from energy-efficiency improvements (e.g. more SUVs), as well as any decrease in average vehicle load factor (e.g. less car sharing). If energy efficiency was measured instead as tonne kilometres per litre of fuel, rebound effects could show up as an increase in tonne kilometres driven, which may be decomposed into the product of the number of vehicles, the mean vehicle weight and the mean distance travelled per vehicle per year. To the extent that vehicle weight provides a proxy for factors such as comfort, safety and carrying capacity, this approach effectively incorporates some features normally classified as attributes of the energy service into the measure of useful work. It also helps to highlight the fact that potential fuel savings from improved engine efficiency may in part be 'taken back' by driving larger cars.

The magnitude of the direct rebound effect may be expected to be proportional to the share of energy in the generalised cost of the useful work (P_S/P_G), as well as the extent to which those costs are 'visible' to the consumer. But as the consumption of a particular energy service increases, saturation effects (technically, declining marginal utility) should reduce the size of any direct rebound effect. For example, direct rebound effects from improvements in the energy efficiency of household heating systems should decline rapidly once whole-house indoor temperatures approach the maximum level for thermal comfort. One important implication is that direct rebound effects will be higher among low income groups, since these are further from satiation in their consumption of many energy services (Boardman and Milne, 2000).

Increases in demand for an energy service may derive from existing consumers of the service, or from consumers who were previously unable or unwilling to purchase that service. For example, improvements in the energy efficiency of home air-conditioners may encourage consumers to purchase portable air-conditioners for the first time. The abundance of such 'marginal consumers' (Wirl, 1997) in developing countries points to the possibility of large rebounds in these contexts, offset to only a limited extent by saturation effects among existing consumers (Roy, 2000).

While energy-efficiency improvements reduce the energy cost of useful work, the size of the direct rebound effect will depend in part upon how such improvements affect other costs. For example, direct rebound effects may be smaller if energy-efficient equipment is more expensive than less efficient alternatives, because the potential saving in operating costs will be offset by higher capital costs. Since consumers typically have high implicit discount rates, the availability of such equipment should not encourage an increase in the number of units purchased, or their average size. However, once purchased, such equipment may be expected to have a higher utilisation. In practice, many types of equipment appear to have both improved in energy efficiency over time and fallen in total cost relative to income.

Even if energy-efficiency improvements are not associated with changes in capital or other costs, certain types of direct rebound effect may be constrained by the real or opportunity costs associated with increasing demand. Two examples are the opportunity cost of space (e.g. increasing refrigerator size may not be the best use of available space) and the opportunity cost of time (e.g. driving longer distances may not be the best use of available time). However, space constraints may become less important over time if technological improvements reduce the average size of conversion devices per unit of output or if rising incomes lead to an increase in average living space (Wilson and Boehland, 2005). In contrast, the opportunity cost of time should increase with rising incomes. The direct rebound effect for a particular energy service may therefore be expected to vary over time and to be influenced by a number of variables.

The quasi-experimental approach to estimating direct rebound effects

One approach to estimating direct rebound effects relies upon measuring the demand for useful work before and after an energy-efficiency improvement: e.g. measuring the change in heat output following the installation of a fuel-efficient boiler. The demand for useful work before the energy-efficiency improvement could be taken as an estimate for what demand 'would have been' in the absence of the improvement. However, various other factors may also have changed the demand for useful work which need to be controlled for (Meyer, 1995; Frondel and Schmidt, 2005).

Since it can be very difficult to measure useful work for many energy services, an alternative approach is to measure the change in energy consumption for that service (e.g. the fuel consumed by the boiler).

But to estimate direct rebound effects, this needs to be compared with a counterfactual estimate of energy consumption that has at least two sources of error, namely: (a) the energy consumption that would have occurred without the energy-efficiency improvement; and (b) the energy consumption that would have occurred following the energy-efficiency improvement had there been no behavioural change. The first of these gives an estimate of the energy savings from the energy-efficiency improvement, while the second isolates the direct rebound effect. Estimates for the latter can be derived from engineering models, but these frequently require data on the circumstances of individual installations and are prone to error.

Both of these approaches are rare, owing in part to measurement difficulties. There are relatively few published studies and nearly all of these focus on household heating (Sommerville and Sorrell, 2007).[1] The methodological quality of most of these studies is relatively poor, with the majority using simple before and after comparisons, without the use of a control group or explicitly controlling for confounding variables. This is the weakest methodological strategy and prone to bias (Meyer, 1995; Frondel and Schmidt, 2005). Also, several studies are vulnerable to selection bias, since households choose to participate rather than being randomly assigned (Hartman, 1988). Other weaknesses include small sample sizes, a failure to present the error associated with estimates, large variation in the relevant independent variable both within and between studies (e.g. households receiving different types of energy efficiency measure, or combination of measures) and monitoring periods that are too short to capture long-term effects. There is also persistent confusion between:

- *shortfall*, the difference between actual savings in energy consumption and those expected on the basis of engineering estimates;
- *temperature take-back*, the change in mean internal temperatures following the energy-efficiency improvement; and
- *behavioural change*, the proportion of the change in internal temperature that derives from adjustments of heating controls and other variables by the user (e.g. opening windows).

Unfortunately, different studies use different terms for the above concepts as well as the same term for different concepts. Typically, only a portion of temperature take-back is due to behavioural change, with the remainder being due to physical and other factors (Sanders and Phillipson, 2006).[2] Similarly, only a portion of shortfall is due to temperature take-back, with the remainder being due to poor engineering estimates of potential savings, inadequate performance of equipment, deficiencies in installation and so on. Hence, behavioural change is one, but not the only (or necessarily the most important) explanation of temperature take-back and the latter is one, but not the only explanation of shortfall. Direct rebound effects are normally interpreted as behavioural change, but it may be misleading to interpret this solely as a rational response to lower heating costs, partly because energy-efficiency improvements may change other variables (e.g. airflow) that also encourage behavioural responses. Also, measures of temperature take-back may be difficult to translate into estimates of shortfall because of the non-linear and context-specific relationship between energy consumption and internal temperature. Isolating direct rebound effects from such studies can therefore be challenging.

The econometric approach to estimating direct rebound effects

A more common approach to estimating direct rebound effects is through the econometric analysis of secondary data sources that include information on the demand for energy, useful work and/or energy efficiency. These data can take a number of forms (e.g. cross-sectional, time-series, panel) and apply to different levels of aggregation (e.g. household, region, country). Such studies typically estimate *elasticities*, meaning the percentage change in one variable following a percentage change in another, holding other variables constant. If time-series data are available, an estimate can be made of short-run elasticities, where the stock of conversion devices is assumed to be fixed, as well as long-run elasticities where it is variable. Cross-sectional data are usually assumed to provide estimates of long-run elasticities.

Depending upon data availability, the direct rebound effect may be estimated from one of two *energy efficiency* elasticities:[3]

- $\eta_\varepsilon(E)$: the elasticity of demand for energy (E) with respect to energy efficiency (ε)
- $\eta_\varepsilon(S)$: the elasticity of demand for useful work (S) with respect to energy efficiency (where $S = \varepsilon E$)

$\eta_\varepsilon(S)$ is generally taken as a direct measure of the rebound effect. Under certain assumptions, it can be shown that: $\eta_\varepsilon(E) = \eta_\varepsilon(S) - 1$ (Sorrell and Dimitropoulos, 2007a). Hence, the actual saving in energy consumption will only equal the predicted saving from engineering calculations when the demand for useful work remains unchanged following an energy-efficiency improvement (i.e. when $\eta_{\varepsilon}(S)=0$). In these circumstances, an x per cent improvement in energy efficiency should lead to an x per cent reduction in energy consumption (i.e. $\eta_{\varepsilon}(E)=-1$).[4]

But instead of using $\eta_{\varepsilon}(E)$ or $\eta_{\varepsilon}(S)$, most studies estimate the rebound effect from one of three *price* elasticities:

- $\eta_{P_S}(S)$: the elasticity of demand for useful work with respect to the energy cost of useful work (P_S)
- $\eta_{P_E}(S)$: the elasticity of demand for useful work with respect to the price of energy (P_E)
- $\eta_{P_E}(E)$: the elasticity of demand for energy with respect to the price of energy

where $P_S=P_E/\varepsilon$. Under certain assumptions, the negative of either $\eta_{P_S}(S)$, $\eta_{P_E}(S)$ or $\eta_{P_E}(E)$ can be taken as an approximation to $\eta_{\varepsilon}(S)$ and hence may be used as a measure of the direct rebound effect (Sorrell and Dimitropoulos, 2007a). The use of price elasticities in this way implicitly equates the direct rebound effect to a behavioural response to the lower cost of useful work. It therefore ignores any other reasons why the demand for useful work may change following an improvement in energy efficiency.

The choice of the appropriate elasticity measure will depend in part upon data availability.[5] Generally, data on energy consumption (E) and energy prices (P_E) are both more available and more accurate than data on useful work (S) and energy efficiency (ε). Also, even if data on energy efficiency are available, the amount of variation is typically limited, with the result that estimates of either $\eta_{\varepsilon}(E)$ or $\eta_{\varepsilon}(S)$ can have a large variance. In contrast, estimates of $\eta_{P_S}(S)$ may have less variance owing to significantly greater variation in the independent variable (P_S). This is because the energy cost of useful work depends upon the ratio of energy prices to energy efficiency ($P_S=P_E/\varepsilon$) and most data sets include considerable cross-sectional or longitudinal variation in energy prices. In principle, rational consumers should respond in the same way to a decrease in energy prices as they do to an improvement in energy efficiency (and vice versa), since these should have an identical effect on the energy cost of useful work (P_S). However, there may be a number of reasons why this 'symmetry' assumption does not hold. If so, estimates of the direct rebound effect that are based upon $\eta_{P_S}(S)$ could be biased.

Estimates of $\eta_{P_S}(S)$ are largely confined to personal automotive transportation, household heating and space cooling, where proxy measures of useful work are most readily available. These energy services form a significant component of household energy consumption in OECD countries and demand for them may be expected to be relatively price responsive. There are very few estimates of $\eta_{P_S}(S)$ for other consumer energy services and practically none for producers. Furthermore, the great majority of studies refer to the United States.

In many cases, data on energy efficiency are either unavailable or inaccurate. In these circumstances, the direct rebound effect may be estimated from $\eta_{P_E}(S)$ and $\eta_{P_E}(E)$. But this is only valid if: first, consumers respond in the same way to a decrease in energy prices as they do to an improvement in energy efficiency (and vice versa); and second, energy efficiency is unaffected by changes in energy prices. Both these assumptions are likely to be flawed, but the extent to which this leads to biased estimates of the direct rebound effect may vary widely from one energy service to another and between the short and long term.

Under certain assumptions, the own-price elasticity of energy demand ($\eta_{P_E}(E)$) for a particular energy service can be shown to provide an *upper bound* for the direct rebound effect.[6] As a result, the voluminous literature on energy price elasticities may be used to place some bounds on the likely magnitude of the direct rebound effect for different energy services in different sectors. This was the approach taken by Khazzoom (1980), who pointed to evidence that the long-run own-price elasticity of energy demand for water heating, space heating and cooking exceeded (minus) unity in some circumstances, implying that energy-efficiency improvements for these services could lead to backfire (Taylor et al., 1977). However, reviews of this literature generally suggest that energy demand is inelastic in the majority of sectors in OECD countries (i.e. $|\eta_{P_E}(E)| < 1$) (Dahl and Sterner, 1991; Dahl, 1993, 1994; Espey, 1998; Graham and Glaister, 2002; Hanley et al., 2002; Espey and Espey, 2004). The implication is that the direct rebound effect alone is unlikely to lead to backfire within OECD countries – although there are undoubtedly exceptions.

For the purpose of estimating rebound effects, estimates of $\eta_{P_E}(E)$ are most useful when the energy demand in question relates to a single energy service, such as refrigeration. They are less useful when (as is more common) the measured demand derives from a collection of energy services, such as household fuel or electricity consumption. In this case, a large value for $\eta_{P_E}(E)$ may suggest that improvements in the 'overall' efficiency of fuel or electricity use will lead to large direct rebound effects (and vice versa), or that the direct rebound effect for the energy services that dominate fuel or electricity consumption may be large. However, a small value for $\eta_{P_E}(E)$ would not rule out the possibility of large direct rebound effects for individual energy services.

Whatever their scope and origin, estimates of price elasticities should be treated with caution. Aside from the difficulties of estimation, behavioural responses are contingent upon technical, institutional, policy and demographic factors that vary widely between different groups and over time. Demand responses are known to vary with the level of prices, the origin of price changes (e.g. exogenous versus policy induced), expectations of future prices, government fiscal policy (e.g. recycling of carbon tax revenues), saturation effects and other factors (Barker et al., 1995). The past is not necessarily a good guide to the future in this area, and it is possible that the very long-run response to price changes may exceed those found in empirical studies that rely upon data from relatively short time periods.

Estimates of direct rebound effects

By far the best-studied area for the direct rebound effect is personal automotive transport. Most studies refer to the US, which is important since fuel prices, fuel efficiencies and residential densities are lower than in Europe, car ownership levels are higher and there is less scope for switching to alternative transport modes.

Studies estimating $\eta_{\varepsilon}(E)$, $\eta_{\varepsilon}(S)$ or $\eta_{P_S}(S)$ for personal transport vary considerably in terms of the data used and specifications employed. Most studies use aggregate data which can capture long-term effects on demand such as fuel efficiency standards, while household survey data can better describe individual behaviour at the micro level. Aggregate studies face numerous measurement difficulties, however (Schipper et al., 1993), while disaggregate studies produce results that are more difficult to generalise. While all studies use vehicle kilometres as the measure of useful work, this may either be measured in absolute terms or normalised to the number of adults, licensed drivers, households or vehicles (Sorrell and Dimitropoulos, 2007b). The relevant estimates may be expected to differ as a result.

Following a review of seventeen econometric studies of personal automotive transport, Sorrell and Dimitropoulos (2007b) conclude that the long-run direct rebound effect is likely to lie somewhere between *10 and 30 per cent*. The most reliable estimates come from studies using cross-country or cross-regional panel data, owing to the greater number of observations. For example, Johansson and Schipper's (1997) cross-country study gives a best guess for the long-run direct rebound effect of 30 per cent, while both Haughton and Sarkar (1996) and Small and Van Dender (2005) converge on a long-run value of 22 per cent for the US (see Box 2.1). Studies using household survey data sources provide less consistent estimates and several of these estimates are higher than those from aggregate data sources. For example, three US studies use data from the same source but produce estimates of the direct rebound effect that range from 0 to 87 per cent (Goldberg, 1996; Puller and Greening, 1999; West, 2004).[7] This diversity suggests that the results from disaggregate studies should be interpreted with more caution. One of the most rigorous studies using disaggregate data is by Greene et al. (1999), who estimate the US average long-run direct rebound effect to be 23 per cent – consistent with the results of studies using aggregate data.

Despite wide differences in data and methodologies, most of these studies provide estimates in the range 10–30 per cent – which suggests that the findings are relatively robust. Moreover, most studies assume that the response to a change in fuel prices is equal in size to the response to a change in fuel efficiency, but opposite in sign. Few studies test this assumption explicitly and those that do are either unable to reject the hypothesis that the two elasticities are equal in magnitude, or find that the fuel efficiency elasticity is *less* than the fuel cost per kilometre elasticity (i.e. $|\eta_{\varepsilon}(S)| < |\eta_{P_E}(S)|$). The implication is that the direct rebound effect may lie towards the lower end of the above range (i.e. around 10 per cent).

A number of studies suggest that the direct rebound effect for personal automotive travel declines with income, as theory predicts (Greene et al., 1999; Small and Van Dender, 2005). The evidence is insufficient to determine whether direct rebound effects are larger or smaller in Europe, but it is notable that the meta-analysis by Espey (1998) found no significant difference in long-run own-price elasticities of gasoline demand ($\eta_{P_E}(E)$). Overall, Sorrell and Dimitropoulos (2007b) conclude that direct rebound effects in this sector have not obviated the benefits of technical improvements in vehicle fuel efficiency at least over the time periods considered. Between 70 and 100 per cent of the potential benefits of such improvements appear to have been realised in reduced consumption of motor-fuels.

Small and Van Dender (2005) provide one of the most methodologically rigorous estimates of the direct rebound effect for personal automotive transport. They estimate an econometric model explaining the amount of travel by passenger cars as a function of the cost per mile and other variables. By employing simultaneous equations for vehicle numbers, average fuel efficiency and vehicle miles travelled, they are able to allow for the fact that fuel efficiency is endogenous: i.e. more fuel-efficient cars may encourage more driving, while the expectation of more driving may encourage the purchase of more fuel-efficient cars. Their results show that failing to allow for this can lead the direct rebound effect to be overestimated.

Box 2.1 The declining direct rebound effect

Small and Van Dender (2005) use aggregate data on vehicle numbers, fuel efficiency, gasoline consumption, vehicle miles travelled and other variables for 50 US states and the District of Columbia covering the period 1961–2001. This approach provides considerably more observations than conventional aggregate time-series data, while at the same time providing more information on effects that are of interest to policy-makers than do studies using household survey data.

Small and Van Dender estimate the short-run direct rebound effect for the US as a whole to be 4.5 per cent and the long-run effect to be 22 per cent. The former is lower than most of the estimates in the literature, while the latter is close to the consensus. However, they estimate that a 10 per cent increase in income reduces the short-run direct rebound effect by 0.58 per cent. Using US average values of income, urbanisation and fuel prices over the period 1997–2001, they find a direct rebound effect of only 2.2 per cent in the short-term and 10.7 per cent in the long-term – approximately half the values estimated from the full data set. If this result is robust, it has some important implications. However, two-fifths of the estimated reduction in the rebound effect derives from the assumption that the magnitude of this effect depends upon the absolute level of fuel costs per kilometre. But since the relevant coefficient is not statistically significant, this claim is questionable.

Although methodologically sophisticated, the study is not without its problems. Despite covering 50 states over a period of 36 years, the data provide relatively little variation in vehicle fuel efficiency making it difficult to determine its effect separately from that of fuel prices. Direct estimates of $\eta_{\varepsilon}(S)$ are small and statistically insignificant, which could be interpreted as implying that the direct rebound effect is approximately zero, but since this specification performs rather poorly overall, estimates based upon $\eta_{P_S}(S)$ are preferred. Also, the model leads to the unlikely result that the direct rebound effect is negative in some states. This raises questions about the use of the model for projecting declining rebound effects in the future, since increasing incomes could make the estimated direct rebound effect negative in many states (Harrison et al., 2005).

The next best-studied area for direct rebound effects is household heating. Sommerville and Sorrell (2007) review fifteen quasi-experimental studies of this energy service and conclude that standard engineering models may overestimate energy savings by up to one half – and potentially by more than this for low income households. However, overall percentage shortfall is highly contingent on the accuracy of the engineering models, and attempts to calibrate these models to specific household conditions generally result in a lower shortfall. The studies reviewed use a wide range of variables to explain shortfall, but only initial energy consumption and the age of the home consistently influence the extent to which predicted savings are likely to be achieved.

The studies provide estimates of temperature take-back in the range 1.0°C to 1.6°C, of which approximately half is estimated to be accounted for by the physical characteristics of the house and the remainder by behavioural change. This behavioural change is not trivial: depending upon insulation standards and external temperatures a 1°C increase in internal temperature may increase the energy consumption for space heating by 10 per cent or more. Temperature take-back appears to be higher for low income groups and for households with low internal temperatures prior to the efficiency measures. However, these two explanatory variables are likely to be correlated. As pre-intervention room temperatures approach 21°C the magnitudes of temperature take-back decrease owing to saturation effects.

Overall, while shortfall may often exceed 50 per cent (especially for low-income households), temperature take-back only accounts for a portion of this shortfall and behavioural change only accounts for a portion of the take-back. Temperature take-back would appear to reduce energy savings by around *20 per cent* on average, with the contribution from behavioural change being somewhat less. Which of these measures best corresponds to the direct rebound effect is a matter of debate. Temperature take-back may be expected to decrease over time as average internal temperatures increase.

Relatively few econometric studies estimate $\eta_{\varepsilon}(E)$, $\eta_{\varepsilon}(S)$ or $\eta_{P_S}(S)$ for household heating and even fewer investigate rebound effects. Most studies rely upon detailed household survey data and exhibit considerable diversity in terms of the variables measured and the methodologies adopted. Sorrell and Dimitropoulos (2007b) review nine studies and find values in the range *10–58 per cent* for the short run and *1.4–60 per cent* for the long run.

As an illustration, Schwarz and Taylor (1995) use cross-sectional data from 1,188 single family US households, including measurements of thermostat settings. They estimate an equation for the thermostat setting as a function of energy prices, external temperature, heated area, household income and an engineering estimate of the thermal resistance of the house. Their data allow them to estimate the demand for useful work for space heating and hence to estimate $\eta_\varepsilon(S)$. Their results suggest a long-run direct rebound effect of between 1.4 and 3.4 per cent.

Contrasting results are obtained by Hseuh and Gerner (1993), who use comparable data from 1,281 single family detached households in the US, dating back to 1981. Their data set includes comprehensive information on appliance ownership and demographic characteristics, which allows them to combine econometric and engineering models to estimate the energy use for space heating. On the basis of an estimate of $\eta_{P_E}(E)$, conditional on the existing level of energy efficiency, the short-run direct rebound effect is estimated as 35 per cent for electrically heated homes and 58 per cent for gas heated homes.

As these examples demonstrate, the definition of the direct rebound effect is not consistent between studies and the behavioural response appears to vary widely between different households. Nevertheless, the econometric evidence broadly supports the conclusions of the quasi-experimental studies, suggesting a mean value for the direct rebound effect for household heating of around *30 per cent.*

Sorrell and Dimitropoulos (2007b) found only two studies of direct rebound effects for household cooling and these provided estimates comparable to those for household heating (i.e. *1–26 per cent*). However, these were relatively old studies, conducted during periods of rising energy prices and using small sample sizes. Their results may not be transferable to other geographical areas, owing to differences in house types and climatological conditions. Also, both studies focused solely upon changes in equipment utilisation. To the extent that ownership of cooling technology is rapidly increasing in many countries, demand from 'marginal consumers' may be an important consideration, together with increases in system capacity among existing users.

Sorrell and Dimitropoulos (2007b) find that the evidence for water heating is even more limited, although Guertin et al. (2003) provide estimates in the range 34–38 per cent, which is significantly larger than the results from quasi-experimental studies reported by Nadel (1993). A methodologically rigorous study of direct rebound effects for clothes washing (Box 2.2) suggests that direct rebound effects for 'minor' energy services should be relatively small (i.e. <5 per cent), as theory suggests. However, this study confines attention to households that already have automatic washing machines and therefore excludes rebound effects from marginal consumers.

Box 2.2 Direct rebound effects for clothes washing

Davis (2007) provides a unique example of an estimate of direct rebound effects for household clothes washing – which together with clothes drying accounts for around one tenth of US household energy consumption. The estimate is based upon a government-sponsored field trial of high-efficiency washing machines involving 98 participants. These machines use 48 per cent less energy per wash than standard machines and 41 per cent less water.

While participation in the trial was voluntary, both the utilisation of existing machines and the associated consumption of energy and water were monitored for a period of two months prior to the installation of the new machine. This allowed household-specific variations in utilisation patterns to be controlled for and permitted unbiased estimates to be made of the price elasticity of machine utilisation.

The monitoring allowed the marginal cost of clothes washing for each household to be estimated. This was then used as the primary independent variable in an equation for the demand for clean clothes in kg/day (useful work). Davis found that the demand for clean clothes increased by 5.6 per cent after receiving the new washers, largely as a result of increases in the weight of clothes washed per cycle rather than the number of cycles. While this could be used as an estimate of the direct rebound effect, it results in part from savings in water and detergent costs. If the estimate was based solely on the savings in energy costs, the estimated effect would be smaller. This suggests that only a small portion of the gains from energy-efficient washing machines will be offset by increased utilisation.

Davis estimates that time costs form 80–90 per cent of the total cost of washing clothes. The results therefore support the theoretical prediction that, for time-intensive activities, even relatively large changes in energy efficiency should have little impact on demand (Binswanger, 2001). Similar conclusions should therefore apply to other time-intensive energy services that are both produced and consumed by households, including those provided by dishwashers, vacuum cleaners, televisions, power tools, computers and printers.

Table 2.1 summarises the results of Sorrell and Dimitropoulos' survey of econometric estimates of the direct rebound effect. Despite the methodological diversity, the results for individual energy services are broadly comparable, suggesting that the evidence is relatively robust to

Table 2.1 Econometric estimates of the long-run direct rebound effect for consumer energy services in the OECD

End-use	*Range of values in evidence base (%)*	*'Best guess' (%)*	*Number of studies*	*Degree of confidence*
Personal automotive transport	3–87	10–30	17	High
Space heating	0.6–60	10–30	9	Medium
Space cooling	1–26	1–26	2	Low
Other consumer energy services	0–41	<20	3	Low

different data sets and methodologies. Also, consideration of the potential sources of bias (see below) suggests that direct rebound effects are more likely to lie towards the lower end of the range indicated here. The results suggest that the mean long-run direct rebound effect for personal automotive transport, household heating and household cooling in OECD countries is likely to be 30 per cent or less and may be expected to decline in the future as demand saturates and income increases. Both theoretical considerations and the limited empirical evidence suggest that direct rebound effects are significantly smaller for other consumer energy services. However, the same conclusion may not follow for energy-efficiency improvements by producers or for low income households in developing countries. Moreover, the evidence base is sparse and has a number of important limitations, including the neglect of marginal consumers, the relatively limited time periods over which most of the effects have been studied and the restricted definitions of 'useful work' that have been employed. For these and other reasons, it would be inappropriate to draw conclusions about rebound effects 'as a whole' from this evidence.

Sources of bias in estimates of the direct rebound effect

Most estimates of the direct rebound effect assume that the change in demand following a change in energy prices is equal to that following a change in energy efficiency, but opposite in sign. Most studies also assume that any change in energy efficiency derives solely from outside the model (i.e. energy efficiency is 'exogenous'). In practice, both of these assumptions may be incorrect.

First, while changes in energy prices are generally not correlated with changes in other input costs, changes in energy efficiency may be.

In particular, higher energy efficiency may only be achieved through the purchase of new equipment with higher capital costs than less efficient models. Hence, estimates of the direct rebound effect that rely primarily upon historical and/or cross-sectional variations in energy prices could overestimate the direct rebound effect, since the additional capital costs required to improve energy efficiency will not be taken into account (Henly et al., 1988).

Second, energy price elasticities tend to be higher for periods with rising prices than for those with falling prices (Gately 1992, 1993; Dargay and Gately, 1994, 1995; Haas and Schipper, 1998). For example, Dargay (1992) found that the reduction in UK energy demand following the price rises of the late 1970s was five times greater than the increase in demand following the price collapse of the mid-1980s. An explanation may be that higher energy prices induce technological improvements in energy efficiency, which also become embodied in regulations (Grubb, 1995). Also, investment in measures such as thermal insulation is largely irreversible over the short to medium term. But the appropriate proxy for improvements in energy efficiency is *reductions* in energy prices. Since many studies based upon time-series data incorporate periods of rising energy prices, the estimated price elasticities may overestimate the response to falling energy prices. As a result, such studies could overestimate the direct rebound effect.

Third, while improved energy efficiency may increase the demand for useful work (e.g. you could drive further after purchasing an energy-efficient car), it is also possible that the anticipated high demand for useful work may increase the demand for energy efficiency (e.g. you purchase an energy-efficient car because you expect to drive further). In these circumstances, the demand for useful work depends on the price of useful work, which depends upon energy efficiency, which depends upon the demand for useful work (Small and Van Dender, 2005). Hence, the direct rebound effect would not be the only explanation for any measured correlation between energy efficiency and the demand for useful work. This so-called 'endogeneity' can be addressed through the use of simultaneous equation models, but these are relatively uncommon owing to their greater data requirements. If, instead, studies use a single equation without the use of appropriate techniques, the resulting estimates could be biased. Several studies of direct rebound effects could be flawed for this reason.

Finally, consumers may be expected to take the full costs of energy services into account when making decisions about the consumption of those services and these include the *time costs* associated with producing and/or using the relevant service – for example, the time required to travel from A to B. Indeed, the increase in energy consumption in industrial societies over the past century may have been driven in part by attempts to 'save time' (and hence time costs) through the use of technologies that allow tasks to be completed faster at the expense of using more energy. For example, travel by private car has replaced walking, cycling and public transport; automatic washing machines have replaced washing by hand; fast food and ready meals have replaced traditional cooking and so on. While not all energy services involve such trade-offs, many do (compare rail and air travel for example). Time costs may be approximated by hourly wage rates and since these have risen more rapidly than energy prices throughout the last century, there has been a strong incentive to substitute energy for time (Becker, 1965). If time costs continue to increase in importance relative to energy costs, the direct rebound effect for many energy services should become *less* important – since improvements in energy efficiency will have an increasingly small impact on the total cost of useful work (Binswanger, 2001). This suggests that estimates of the direct rebound effect that do not control for increases in time costs (which is correlated with increases in income) could potentially overestimate the direct rebound effect. Box 2.1 shows how this could be particularly relevant to direct rebound effects in transport. Similar reasoning suggests that the direct rebound effect may decline as the mean level of energy-efficiency improves as energy costs should form a declining fraction of the total cost of useful work.

The consideration of time costs also points to an important but relatively unexplored issue: increasing time efficiency may lead to a parallel 'rebound effect with respect to time' (Binswanger, 2001; Jalas, 2002). For example, faster modes of transport may encourage longer commuting distances, with the time spent commuting remaining broadly unchanged. So in some circumstances energy consumption may be increased, first, by trading off energy efficiency for time efficiency (e.g. choosing air travel rather than rail) and second, by the rebound effects with respect to time (e.g. choosing to travel further).

Summary

In summary, the accurate estimation of direct rebound effects is far from straightforward. A prerequisite is adequate data on energy consumption, useful work and/or energy efficiency which are only available for a subset of energy services. As a consequence, the evidence remains sparse, inconsistent and methodologically diverse, as well as being largely confined to a limited number of consumer energy services in the United States. It is important to recognise that estimates of the direct rebound effect for these energy services provide no guide for the magnitude of such effects in other sectors, or for the rebound effect overall.

The evidence for direct rebound effects for automotive transport and household heating is relatively robust while that for other consumer energy services is much weaker. Evidence is particularly weak for energy-efficiency improvements in developing countries, although theoretical considerations suggest that direct rebound effects in this context could be much larger.

Under certain assumptions, estimates of the own-price elasticity of energy demand for an individual energy service should provide an upper bound for the direct rebound effect for that service. If the measured energy demand relates to a group of energy services (e.g. household fuel demand), the own-price elasticity should provide an approximate upper bound for the weighted average of direct rebound effects for those services. Since the demand for energy is generally found to be inelastic in OECD countries, the long-run direct rebound effect for most energy services should be less than 100 per cent.

For personal automotive transport, household heating and household cooling in OECD countries, the mean value of the long-run direct rebound effect is likely to be less than 30 per cent and may be closer to 10 per cent for transport. Moreover, the effect is expected to decline in the future as demand saturates and income increases. Both theoretical considerations and the available empirical evidence suggest that direct rebound effects should be smaller for other consumer energy services where energy forms a small proportion of total costs. Hence, at least for OECD countries, direct rebound effects should only partially offset the energy savings from energy-efficiency improvements in consumer energy services.

These conclusions are subject to a number of qualifications, including the relatively limited time periods over which direct rebound effects have been studied and the restrictive definitions of 'useful work' that have been employed. For example, current studies only measure the increase in distance driven for automotive transport and do not measure changes in vehicle size. Rebound effects for space heating and other energy services are also higher among low income groups and most studies do not account for 'marginal consumers' acquiring services such as space cooling for the first time.

The methodological quality of many quasi-experimental studies is poor, while the estimates from many econometric studies appear vulnerable to bias. The most likely effect of the latter is to lead the direct rebound effect to be overestimated. Considerable scope exists for improving estimates of the direct rebound effect for the energy services studied here and for extending estimates to include other energy services. But this can only be achieved with better data.

Acknowledgements

This chapter is based upon a comprehensive review of the evidence for rebound effects, conducted by the UK Energy Research Centre (Sorrell, 2007). An earlier version of this chapter is contained in Sorrell (2007), while the research upon which it is based is reported in detail in Sommerville and Sorrell (2007) and Sorrell and Dimitropoulos (2007b). The financial support of the UK Research Councils is gratefully acknowledged.

Considerable thanks are due to Matt Sommerville (Imperial College) and John Dimitropoulos (Sussex Energy Group) for their contribution to this work. The author is also grateful for the advice and comments received from Manuel Frondel, Karsten Neuhoff, Jake Chapman, Nick Eyre, Blake Alcott, Horace Herring, Paolo Agnolucci, Jim Skea, Rob Gross and Phil Heptonstall. A debt is also owed to Lorna Greening and David Greene for their previous synthesis of empirical work in this area (Greening and Greene, 1998). The usual disclaimers apply.

Notes

1. Nadel (1993) reports the results of a number of quasi-experimental studies by US utilities, which suggest direct rebound effects of 10 per cent or less for lighting and approximately zero for water heating, with inconclusive results for refrigeration. Sommerville and Sorrell (2007) were not able to access these studies, which appear to be small-scale, short-term and methodologically weak. Instead, they summarise the results of fifteen studies of household heating.
2. For example, daily average household temperatures will generally increase following improvements in thermal insulation, even if the heating controls remain unchanged. This is because insulation contributes to a more even distribution of warmth around the house, reduces the rate at which a house cools down when the heating is off and delays the time at which it needs to be switched back on (Milne and Boardman, 2000).
3. The rationale for the use of these elasticities, and the relationship between them, is explained in detail in Sorrell and Dimitriopoulos (2007b). In particular, it is shown that the following relationship may be expected to hold: $|\eta_{P_E}(S)| \leq |\eta_{P_S}(S)| \leq |\eta_{P_E}(E)| \leq |\eta_{P_S}(E)|$.
4. A positive rebound effect implies that $\eta_\varepsilon(S) > 0$ and $0 > \eta_\varepsilon(S) > -1$, while backfire implies that $\eta_\varepsilon(S) > 1$ and $\eta_\varepsilon(E) > 0$.

5. In the case of personal automotive transport, for example, the elasticities could correspond to: $\eta_{\varepsilon}(E)$ – the elasticity of the demand for motor-fuel (for passenger cars) with respect to kilometres per litre; $\eta_{\varepsilon}(S)$ – the elasticity of the demand for vehicle kilometres with respect to kilometres per litre; $\eta_{P_S}(S)$ – the elasticity of the demand for vehicle kilometres with respect to the cost per kilometre; $\eta_{P_E}(S)$ – the elasticity of the demand for vehicle kilometres with respect to the price of motor-fuel; and $\eta_{P_E}(E)$ – the elasticity of the demand for motor-fuel with respect to the price of motor-fuel.
6. As an illustration, the upper bound for the direct rebound effect for personal automotive transport can be estimated from Hanley et al.'s (2002) meta-analysis of 51 empirical estimates of the long-run own-price elasticity of motor-fuel demand ($\eta_{P_E}(E)$). Taking the mean values, this suggests an upper bound for the long-run direct rebound effect for this energy service of 64 per cent. For comparison, three estimates of the elasticity of distance travelled by private cars with respect to the price of motor-fuel ($\eta_{P_E}(S)$) suggest a smaller long-run direct rebound effect of only 30 per cent (Hanley et al., 2002). The difference between the two suggests that fuel prices have a significant influence on vehicle fuel efficiency over the long term and shows the extent to which the use of $\eta_{P_E}(E)$ may lead to the direct rebound effect being overestimated.
7. All use the US Consumer Expenditure Survey, although covering different time periods and supplemented by differing sources of information on vehicle fuel efficiency. The figure of 87 per cent is from West (2004) and is likely to be an overestimate of the direct rebound effect since it derives from the estimated elasticity of distance travelled with respect to *operating* costs, which includes maintenance and tyre costs.

References

Barker, T., P. Ekins and N. Johnstone (1995) *Global Warming and Energy Demand*, London: Routledge.

Becker, G. S. (1965) 'A theory of the allocation of time', *Economic Journal*, 75(299): 493–517.

Binswanger, M. (2001) 'Technological progress and sustainable development: what about the rebound effect?' *Ecological Economics*, 36(1): 119–32.

Boardman, B. and G. Milne (2000) 'Making cold homes warmer: the effect of energy-efficiency improvements in low-income homes', *Energy Policy*, 218(6–7): 411–24.

Dahl, C. (1993) *A Survey of Energy Demand Elasticities in Support of the Development of the NEMS*. Prepared for US Department of Energy, Contract No. De-AP01-93EI23499, Department of Mineral Economics, Colorado School of Mines, Colorado.

Dahl, C. (1994) 'Demand for transportation fuels: a survey of demand elasticities and their components', *Journal of Energy Literature*, 1(2): 3.

Dahl, C. and T. Sterner (1991) 'Analyzing gasoline demand elasticities – a survey', *Energy Economics*, 13(3): 203–10.

Dargay, J. M. (1992) *Are Price and Income Elasticities of Demand Constant? The UK Experience*, Oxford: Oxford Institute for Energy Studies.

Dargay, J. M. and D. Gately (1994) 'Oil demand in the industrialised countries', *Energy Journal*, 15: 39–67.
Dargay, J. M. and D. Gately (1995) 'The imperfect price irreversibility of non-transportation of all demand in the OECD', *Energy Economics*, 17(1): 59–71.
Davis, L. W. (2007) *Durable Goods and Residential Demand for Energy and Water: Evidence from a Field Trial*, Working Paper, Department of Economics, University of Michigan.
Department of Transport (2006) *National Travel Survey*, London: Department of Transport.
Espey, J. A. and M. Espey (2004) 'Turning on the lights: a meta-analysis of residential electricity demand elasticities', *Journal of Agricultural and Applied Economics*, 36(1): 65–81.
Espey, M. (1998) 'Gasoline demand revisited: an international meta-analysis of elasticities', *Energy Economics*, 20: 273–95.
Frondel, M. and C. M. Schmidt (2005) 'Evaluating environmental programs: the perspective of modern evaluation research', *Ecological Economics*, 55(4): 515–26.
Gately, D. (1992) 'Imperfect price reversed ability of US gasoline demand: asymmetric responses to price increases and declines', *Energy Journal*, 13(4): 179–207.
Gately, D. (1993) 'The imperfect price reversibility of world oil demand', *Energy Journal*, 14(4): 163–82.
Goldberg, P. K. (1996) *The Effect of the Corporate Average Fuel Efficiency Standards*, Working Paper No. 5673, National Bureau of Economic Research, Cambridge, MA.
Graham, D. J. and S. Glaister (2002) 'The demand for automobile fuel: a survey of elasticities', *Journal of Transport Economics and Policy*, 36(1): 1–26.
Greene, D. L., J. R. Kahn and R. C. Gibson (1999) 'Fuel economy rebound effect for US household vehicles', *Energy Journal*, 20(3): 1–31.
Greening, L. A. and D. L. Greene (1998) *Energy Use, Technical Efficiency, and the Rebound Effect: a Review of the Literature*, Report to the US Department of Energy, Hagler Bailly and Co., Denver.
Grubb, M. J. (1995) 'Asymmetrical price elasticities of energy demand', in T. Barker, P. Ekins and N. Johnstone (eds), *Global Warming and Energy Demand*, London: Routledge.
Guertin, C., S. Kumbhakar and A. Duraiappah (2003) *Determining Demand for Energy Services: Investigating Income-driven Behaviour*, International Institute for Sustainable Development.
Haas, R. and L. Schipper (1998) 'Residential energy demand in OECD-countries and the role of irreversible efficiency improvements', *Energy Economics*, 20(4): 421–42.
Hanley, M., J. M. Dargay and P. B. Goodwin (2002) 'Review of income and price elasticities in the demand for road traffic', Final report to the DTLR under Contract number PPAD 9/65/93, London: ESRC Transport Studies Unit, University College London.
Harrison, D., G. Leonard, B. Reddy, D. Radov, P. Klevnäs, J. Patchett and P. Reschke (2005) *Reviews of Studies Evaluating the Impacts of Motor Vehicle Greenhouse Gas Emissions Regulations in California*, Boston: National Economic Research Associates.

Hartman, R. S. (1988) 'Self-selection bias in the evaluation of voluntary energy conservation programs', *Review of Economics and Statistics*, 70(3): 448–58.

Haughton, J. and S. Sarkar (1996) 'Gasoline tax as a corrective tax: estimates for the United States 1970–1991', *Energy Journal*, 17(2): 103–26.

Henly, J., H. Ruderman and M. D. Levine (1988) 'Energy savings resulting from the adoption of more efficient appliances: a follow-up', *Energy Journal*, 9(2): 163–70.

Hsueh, L.-M. and J. L. Gerner (1993) 'Effect of thermal improvements in housing on residential energy demand', *Journal of Consumer Affairs*, 27(1): 87–105.

Jalas, M. (2002) 'A time use perspective on the materials intensity of consumption', *Ecological Economics*, 41(1): 109–23.

Johansson, O. and L. Schipper (1997) 'Measuring long-run automobile fuel demand: separate estimations of vehicle stock, mean fuel intensity, and mean annual driving distance', *Journal of Transport Economics and Policy*, 31(3): 277–92.

Khazzoom, J. D. (1980) 'Economic implications of mandated efficiency in standards for household appliances', *Energy Journal*, 1(4): 21–40.

Meyer, B. (1995) 'Natural and quasi experiments in economics', *Journal of Business and Economic Statistics*, 13(2): 151–60.

Milne, G. and B. Boardman (2000) 'Making cold homes warmer: the effect of energy-efficiency improvements in low-income homes', *Energy Policy*, 28: 411–24.

Nadel, S. (1993) 'The take-back effect: fact or fiction', U933, ACEEE.

Patterson, M. G. (1996) 'What is energy efficiency? Concepts, indicators and methodological issues', *Energy Policy*, 24(5): 377–90.

Puller, S. L. and L. A. Greening (1999) 'Household adjustment to gasoline price change: an analysis using 9 years of US survey data', *Energy Economics*, 21(1): 37–52.

Roy, J. (2000) 'The rebound effect: some empirical evidence from India', *Energy Policy*, 28(6–7): 433–8.

Sanders, M. and M. Phillipson (2006) 'Review of differences between measured and theoretical energy savings for insulation measures', Glasgow: Centre for Research on Indoor Climate and Health, Glasgow Caledonian University.

Schipper, L., M. Josefina, L. P. Figueroa and M. Espey (1993) 'Mind the gap: the vicious circle of measuring automobile fuel use', *Energy Policy*, 21(12): 1173–90.

Schwarz, P. M. and T. N. Taylor (1995) 'Cold hands, warm hearth? Climate, net takeback, and household comfort', *Energy Journal*, 16(1): 41–54.

Small, K. A. and K. Van Dender (2005) 'A study to evaluate the effect of reduced greenhouse gas emissions on vehicle miles travelled'. Prepared for the State of California Air Resources Board, the California Environment Protection Agency and the California Energy Commission, Final Report ARB Contract Number 02-336, Department of Economics, University of California, Irvine.

Sommerville, M. and S. Sorrell (2007) *UKERC Review of Evidence for the Rebound Effect: Technical Report 1: Evaluation Studies*, London: UK Energy Research Centre.

Sorrell, S. (2007) *The Rebound Effect: an Assessment of the Evidence for Economy-wide Energy Savings from Improved Energy Efficiency*, London: UK Energy Research Centre.

Sorrell, S. and J. Dimitropoulos (2007a) 'The rebound effect: microeconomic definitions, limitations and extensions', *Ecological Economics*, in press.

Sorrell, S. and J. Dimitropoulos (2007b) 'UK ERC review of evidence for the rebound effect: Technical Report 3: Econometric Studies', London: UK Energy Research Centre.

Taylor, L. D., G. R. Blattenberger and P. K. Verleger (1977) *The Residential Demand for Energy*, Report EA-235, Electric Power Research Institute, Palo Alto, California.

West, S. E. (2004) 'Distributional effects of alternative vehicle pollution control policies', *Journal of Public Economics*, 88: 735–57.

Wilson, A. and J. Boehland (2005) 'Small is beautiful: US house size, resource use, and the environment', *Journal of Industrial Ecology*, 9(1–2): 277–87.

Wirl, F. (1997) *The Economics of Conservation Programs*, Dordrecht: Kluwer.

3

Fuel Efficiency and Automobile Travel in Germany: Don't Forget the Rebound Effect!

Manuel Frondel, Jörg Peters and Colin Vance

The improvement of energy efficiency is often asserted to be one of the most promising options to reduce both the usage of energy and associated negative externalities, such as carbon dioxide emissions (CO_2). Ever since the creation of the Corporate Average Fuel Economy (CAFE) standards in 1975, this assertion has been a mainstay of energy policy in the United States. In recent years, it has also found increasing currency in Europe, as attested to by the voluntary agreement negotiated in 1999 between the European Commission (EC) and the European Automobile Manufacturers Association, stipulating the reduction of average emissions to a target level of 140 g CO_2/km by 2008. The EC is additionally considering legislation that would set a target of 120 g CO_2/km by 2012.

Although such technological standards undoubtedly confer benefits via reduced per-unit prices of energy services, the extent to which they reduce energy consumption, and hence pollution, remains controversial. It is plausible, for instance, that the owner of a more fuel-efficient car will *ceteris paribus* drive more in response to lower per-kilometre travelling costs relative to other modes. This increase in service demand from reduced energy prices is called the 'rebound effect', alternatively referred to as 'take-back' of efficiency improvements. Khazzoom (1980) was among the first to study the rebound effect at the microeconomic level of households, focusing on the effects of increases in the energy efficiency of a single energy service, such as space heating and individual conveyance. The 'rebound', however, is a general economic phenomenon, diminishing the potential gains of time-saving technologies (e.g. Binswanger, 2001) as well as of innovations that may reduce the usage of resources such as water.

The significance of the 'rebound' has been hotly debated among energy economists, as documented in several surveys of the relevant literature (e.g. Brookes, 2000; Greening et al., 2000). Though the basic mechanism is widely accepted, the core of the controversy lies in the identification of the magnitude of the *direct rebound effect*, which describes the increased demand for an energy service whose price shrinks due to improved efficiency.[1] This substitution mechanism in favour of the energy service works exactly as would the price reduction of any commodity other than energy, and suggests that price elasticities are at issue when it comes to the estimation of direct rebound effects. Some analysts, most notably Lovins (1988), maintain that these effects are so insignificant that they can safely be ignored (see also Greene, 1992; Schipper and Grubb, 2000). Other authors argue that they might be so large as to completely defeat the purpose of energy-efficiency improvements (Brookes, 1990; Saunders, 1992; Wirl, 1997).

Support for both views is found in the available empirical evidence. In the case of personal automotive transport, for example, a survey by Goodwin et al. (2004) indicates rebound effects – in terms of own-price elasticities of fuel consumption – varying between 4 and 89 per cent from studies using pooled cross-section/time-series data. Results from subsequent studies of this energy service are equally wide-ranging. Using cross-sectional micro data from the 1997 *Consumer Expenditure Survey*, West (2004) finds a rebound effect that is 87 per cent on average, while Small and Van Dender (2007), who use a pooled cross-section of US states for 1966–2001, uncover rebound effects varying between 2.2 and 15.3 per cent.

Aside from differences in the level of data aggregation, one major reason for the diverging results of the empirical studies is that there is no unanimous definition of the direct rebound effect. Instead, several definitions have been employed as determined by the availability of price and efficiency data, making comparisons across studies difficult. The resulting variety of definitions used in the economic literature is summarised and analysed in an illuminating way by Dimitropoulos and Sorrell (2006), who argue that the omission of potentially relevant factors such as capital cost may lead many empirical studies to overestimate the rebound effect. Greene et al. (1999) and Small and Van Dender (2007) express similar reservations, noting in particular the shortcomings of cross-sectional or pooled approaches that fail to control for the time-invariant effects of neighbourhood design, infrastructure and other geographical features, which are likely to be strongly correlated with fuel economy and travel.

Departing from the theoretical grounds provided by Becker's (1965) classical household production function approach and drawing on a panel of household travel data, this chapter focuses on estimating the direct rebound effect from variation in the fuel economy of household vehicles. After cataloguing three commonly employed definitions of the direct rebound effect in the theoretical section of the chapter, econometric estimates corresponding to each of the three definitions are provided in the empirical section. These estimates are generated from panel models of micro-level data, thereby bypassing aggregation problems while at the same time controlling for time-invariant omitted variables.

Our results, which range between 57 and 67 per cent, indicate a 'rebound' that is substantially larger than the typical effects obtained from the US transport sector. Based on household survey data, Greene et al. (1999: 1), for instance, find a long-run 'take-back' of about 20 per cent of potential energy savings, confirming the results of other US studies using national and or state-level data (Sorrell and Dimitropoulos, 2007). While this issue has received relatively less scrutiny in the European context, our results are also substantially larger than those of Walker and Wirl (1993), who estimate a long-run rebound effect of 36 per cent for Germany using aggregate time-series data.

The following section presents three definitions of the direct rebound effect, building the basis for the empirical estimation. Section 3 describes the econometric specifications and estimators. Section 4 describes the panel data base used in the estimation, followed by the presentation and interpretation of the results in Section 5. The last section summarises and concludes.

A variety of direct rebound effect definitions

Following Becker's seminal work on household production, we assume that an individual household derives utility from energy services, such as mobility or comfortable room temperature. A specific service is taken to be the output of a production function f:

$$s = f(e, t, k, o) \tag{1}$$

where f describes how households 'produce' the service in the amount of s by using energy, e, time, t, capital, k, and other market goods o. Using this framework, we begin by drawing on the definition of energy efficiency typically employed in the economic literature (e.g. Wirl, 1997):

$$\mu = \frac{s}{e} > 0, \tag{2}$$

where the efficiency parameter μ characterises the technology with which a service is provided. For the specific example of individual conveyance, parameter μ designates fuel efficiency which can be measured in terms of vehicle kilometres per litre of fuel input. Efficiency definition (2) assumes proportionality between service level and energy input regardless of the level – a simplifying assumption that may not be true in general, but provides for a convenient first-order approximation of the relationship of s with respect to e.

Efficiency definition (2) reflects the fact that the higher the efficiency μ of a given technology, the less energy $e = s/\mu$ is required for the provision of a certain amount s of energy service. Hence, the concept of energy efficiency is perfectly in line with Becker's idea of household production, according to which households are, ultimately, not interested in the energy input required for a certain amount of service, but in the energy service itself. Based on efficiency definition (2), it follows that the price ps per unit of the energy service, given by the ratio of service cost to service amount, is smaller the higher the efficiency μ is:

$$p_s = \frac{e \cdot p_e}{s} = \frac{e}{s} \cdot p_e = \frac{p_e}{\mu}. \tag{3}$$

We now provide a concise summary of three widely known definitions of the *direct* rebound effect that are based on either efficiency, service price, or energy price elasticities. Using these definitions and data on fuel efficiency, fuel prices, distance driven, and fuel consumption for household vehicles originating from German household data, we will estimate each of the three measures of the rebound effect.

Definition 1: The immediate and most general measure of the direct rebound effect (see e.g. Berkhout et al., 2000) is given by $\eta_\mu(s) := \partial \ln s / \partial \ln \mu$, the elasticity of service demand with respect to efficiency, reflecting the relative change in service demand due to a percentage increase in efficiency.

Proposition 1: Having $\eta_\mu(s)$ in hand, we obtain the relative reduction in energy use due to a percentage change of efficiency (see Sorrell and Dimitropoulos, 2008):

$$\eta_\mu(e) = \eta_\mu(s) - 1. \tag{4}$$

Only if $\eta_\mu(s)$ equals zero, that is, only if there is no direct rebound effect, $\eta_\mu(e)$ amounts to -1, indicating that 100 per cent of the potential energy savings due to an efficiency improvement can actually be realised.

Definition 2: Instead of $\eta_\mu(s)$, empirical estimates of the rebound effect are frequently based on $-\eta_{p_s}(s)$, the negative price elasticity of service demand (see e.g. Greene et al., 1999; Binswanger, 2001). Major reasons for this preference are that data on energy efficiency are often unavailable or data provide only limited variation in efficiencies. The basis for this definition is given by the following proposition.

Proposition 2: If energy prices p_e are exogenous and service demand solely depends on p_s, then

$$\eta_\mu(s) = -\eta_{p_s}(s). \tag{5}$$

That the rebound may be captured by $-\eta_{p_s}(s)$ reflects the fact that the direct rebound effect is, in essence, a price effect, which works through shrinking service prices p_s.

Definition 3: Empirical estimates of the rebound effect are sometimes necessarily based on $-\eta_{p_e}(e)$, the negative own-price elasticity of energy consumption, rather than on $-\eta_{p_s}(s)$, because data on energy consumption and prices are more commonly available than on energy services and service prices. It was this definition of the rebound that was originally introduced by Khazzoom (1980: 38) and is also employed by, for example, Wirl (1997: 30). The basis for Definition 3 is given by the following proposition.

Proposition 3: If the energy efficiency μ is constant, then

$$\eta_{p_e}(e) = \eta_{p_s}(s). \tag{6}$$

It bears emphasising that Definitions 2 and 3 are based on the assumption that service demand solely varies with the energy input e, or alternatively is only a function of service price p_s, as is the conventional assumption in the literature. In other words, the possibility that efficiency improvements may not only increase service demand, but also determine other factors, such as the time usage required by an energy service, the use of other commodities, or capital cost, is not considered. In practice, however, more energy-efficient appliances frequently have higher fixed costs while simultaneously reducing operating costs through lower fuel requirements, a point to which we return later on.

Methodology

Our empirical methodology proceeds with two principal aims: (1) to compare alternative model specifications that yield estimates corresponding to each of the three definitions of the rebound effect explicated in the theoretical discussion; (2) to generate these estimates using various panel data estimators that control for the omission of potentially relevant factors varying across observations and over time.

Referring to Definition 1, the first specification regresses the log of monthly kilometres travelled, $\ln(s)$, on the log of kilometres travelled per litre, $\ln(\mu)$, the coefficient of which yields the rebound effect, $\eta_{\mu}(s)$. As control variables, we additionally include the logged price of fuel per litre, $\ln(p_e)$, and a set of household- and car-level variables designated by the vector x.

Model 1:

$$\ln(s_{it}) = \alpha_0 + \alpha_{\mu} \cdot \ln(\mu_{it}) + \alpha_{p_e} \cdot \ln(p_{e_{it}}) + \alpha_x^T \cdot x_{it} + \xi_i + \upsilon_{it}. \tag{7}$$

Subscripts i and t are used to denote the observation and time period, respectively. ξ_i designates an unknown individual-specific term, and υ_{it} is a random component that varies over individuals and time.

The second model generates estimates of the rebound corresponding to Definition 2, which involves regressing $\ln(s)$ on the logged price of fuel per kilometre, $\ln(p_s)$, and the vector of control variables x. In this model, the rebound effect is obtained according to Proposition 2 by the negative coefficient of $\ln(p_s)$:$\eta_{\mu}(s) = -\eta_{p_s}(s) = -\alpha_{p_s}$.

Model 2:

$$\ln(s_{it}) = \alpha_0 + \alpha_{p_s} \cdot \ln(p_{s_{it}}) + \alpha_x^T \cdot x_{it} + \xi_i + \upsilon_{it}. \tag{8}$$

Recognising that $p_s = \frac{p_e}{\mu}$, and that $\ln(p_s) = \ln(p_e) - \ln(\mu)$, it can be seen that the specification of Model 2 is functionally equivalent to that of Model 1. In fact, if we impose the restriction

$$H_0\colon \alpha_{\mu} = -\alpha_{p_e}$$

on Model 1, we exactly get Model 2. Hence, testing the null hypothesis H_0 using Model 1 allows for a simple examination of whether both models are equivalent.

Corresponding to our third definition of the rebound effect, the final specification regresses the logged monthly litres of fuel consumed, $\ln(e)$, on $\ln(p_e)$ and the vector of control variables x.

Model 3:

$$\ln(e_{it}) = \alpha_0 + \alpha_{p_e} \cdot \ln(p_{e_{it}}) + \alpha_x^T \cdot x_{it} + \xi_i + \upsilon_{it}. \tag{9}$$

According to Propositions 2 and 3, the rebound effect results from the negative of the price coefficient: $\eta_\mu(s) = -\eta_{p_e}(e) = -\alpha_{p_e}$. It also bears noting that it is possible to examine whether Model 3 differs from Model 2 by testing the hypothesis

$$H_0 = \alpha_{p_e} = -1 \tag{10}$$

on the basis of the estimates of Model 3 and, additionally, by testing the hypothesis

$$H_0\colon \alpha_{p_s} = -1 \tag{11}$$

on the basis of the estimates of Model 2. Only if both hypotheses were to hold would Model 2 and 3 be identical, as can be seen by inserting restriction (10) into model formulation (9) and employing efficiency definition (2), $\mu = s/e$, and relationship (3), $p_s = p_e/\mu$.

Panel data afford three principal approaches for econometric modelling: the fixed-, between-groups, and random-effects estimators (see the annex for a more detailed, technical exposition of these estimation techniques). While the random-effects estimator is a matrix-weighted average of the fixed- and between-groups effects estimators, the key advantage of the fixed-effects estimator is that it produces consistent estimates even in the presence of time-invariant, unobservable factors ξ_i (e.g. topography and urban form) that vary across households and are correlated with the explanatory variables. In effect, fixed-effects estimations are equivalent to OLS regressions in which dummy variables are typically employed to capture the household-specific factors ξ_i.

In contrast to the fixed-effects estimator, alternatively called the within-groups effects estimator, random effects treats the time-invariant, unobservable factors ξ_i as part of the disturbances, thereby assuming that their correlation with the regressors is zero. If this assumption is met, the random-effects estimator is a viable alternative, as it confers the advantage of greater efficiency over the fixed-effects estimator from its omission of the individual household dummy variables. Violation of the assumption of zero correlation, however, implies biased estimates.

While most analyses neglect between-groups effects, instead focusing on the choice among fixed and random effects, we see merit in applying all three estimators to the three model specifications. For starters, our relatively short panel of three years means that some of the regressors may have insufficient variability to be precisely estimated using fixed effects, a problem that does not afflict between-groups effects given its reliance on cross-sectional information. Beyond this, the between-groups effects estimation, which is equivalent to an OLS regression of averages across time, conveys valuable economic content that is not otherwise revealed. Specifically, while fixed-estimation effects tells us the effect of an explanatory variable as it inter-temporally changes within subjects, between-groups effects tells us the cross-sectional effects of changes in an explanatory variable between subjects.

On the other hand, by averaging household information across time for each variable, the between-groups effects estimator does not fully exploit all the available information. While this deficiency is the reason for the fact that empirical panel studies rarely report between-groups effects, this kind of estimation generally yields long-run effects, as it is based on time averages of household observations. Nevertheless, we acknowledge that the three-year time frame of our data affords limited scope for discerning long-run effects using the between-groups effects estimator.

Last but equally important, we distinguish between fixed and random effects using the classical Hausman test that, in essence, is based on the null hypothesis that the fixed effects are equal to the random effects, which, if not rejected, would suggest adoption of the random-effects estimator due to its higher efficiency. Yet, testing the hypothesis that the fixed and the random effects are equal is numerically identical to testing that the between groups and fixed effects are equal (see e.g. Baltagi, 2005: 67) and thus that the inter-temporal within-groups effects are the same as the cross-sectional effects across subjects. As there is rarely a theoretical basis for this assumption, it must not be surprising if the null hypothesis of the Hausman test is rejected.

The German Mobility Panel data set

The data used in this research are drawn from the German Mobility Panel (MOP, 2007), an ongoing travel survey that was initiated in 1994. The panel is organised in overlapping waves, each comprising a group of households surveyed for a period of one week in autumn for three consecutive years. With each year, a subset of households – those having been surveyed for three years – exits the panel and is replaced by a new group, so that different households are monitored in different years. All households that participate in the survey are requested to fill out a questionnaire eliciting general household information, person-related characteristics, and relevant aspects of everyday travel behaviour. In addition to this general survey, the MOP includes another survey focusing specifically on vehicle travel among a subsample of randomly selected car-owning households. This survey takes place over a roughly six-week period in the spring, during which time respondents record the price paid for fuel, the litres of fuel consumed, and the kilometres driven with each visit to a petrol station and for every car in the household.

The data used in this chapter cover nine years of the survey, spanning 1997 through 2005, a period during which real fuel prices rose 3.2 per cent per annum. To avoid complications of multiple car ownership due to substitution effects among cars, we focus on single-car households, which represent roughly 53 per cent of the households in Germany (MiD, 2007).[2] The resulting sample includes 547 households, 293 of which appear three consecutive years in the data and the remaining 254 of which, due to panel attrition, appear two consecutive years. To correct for the non-independence of repeated observations from the same households, the regression disturbance terms are clustered at the level of the household, and the presented measures of statistical significance are robust to this survey design feature.

We used the travel survey information, which is recorded at the level of the automobile, to derive the dependent and explanatory variables required for estimating each of the three variants of the rebound effect. The two dependent variables, which are converted into monthly figures to adjust for minor variations in the survey duration, are the total monthly distance driven in kilometres (Definitions 1 and 2) and the total monthly litres of fuel consumed (Definition 3). The three explanatory variables for identifying the direct rebound effect are the kilometres travelled per litre (Definition 1), the price paid for fuel per kilometre travelled (Definition 2), and the price paid for fuel per litre (Definition 3).[3]

The remaining suite of variables selected for inclusion in the model measure the socio-demographic and automobile attributes that are hypothesised to influence the extent of motorised travel. Table 3.1 contains the definitions and descriptive statistics of all the variables used in the modelling.

To control for the effects of quality (Wirl, 1997: 14), the age of the automobile and a dummy indicating luxury models are included.[4] Although income is not directly measured, an attempt is made to proxy for its influence via measures of the number of employed residents and the number with a high school diploma living in the household. Finally, controls are included for household size, the presence of children, whether the household undertook a vacation with the car during the survey period,

Table 3.1 Variable definitions and descriptive statistics

Variable definition	*Variable name*	*Mean*	*Std. dev.*
Monthly kilometres driven	s	1156.82	714.55
Monthly fuel consumption in litres	e	94.82	59.33
Kilometres driven per litre	μ	12.51	2.76
Real fuel price in euros per kilometre	p_s	0.08	0.02
Real fuel price in euros per litre	p_e	0.96	0.13
Age of the car	*car age*	6.13	4.04
Dummy: 1 if fuel type is diesel	*diesel car*	0.10	0.30
Dummy: 1 if car is a sports- or luxury model	*premium car*	0.21	0.40
Number of household members	*household size*	1.98	1.04
Number of household members with a high school diploma	*# high school diploma*	0.48	0.65
Number of employed household members	*# employed*	0.71	0.74
Dummy: 1 if household undertook car vacation during the survey period	*car vacation*	0.24	0.43
Dummy: 1 if children younger than 12 live in household	*children*	0.13	0.34
Dummy: 1 if an employed household member changed jobs within the preceding year	*job change*	0.10	0.30

and whether any employed member of the household changed jobs in the preceding year.

Empirical results

Our empirical analysis of the data involved the estimation of two sets of models, one in which the individual effects (ξ_i) were specified at the level of the household and one in which they were specified at the level of the automobile. Noting that this distinction had little bearing on the qualitative conclusions of the analysis, the following discussion focuses on the estimates generated at the household level. This focus facilitates comparison of the three estimators as it ensures that each uses the same sample of observations. Were the individual component set at the level of the automobile, then observations in which the household changes automobiles from one year to the next would drop out in the case of the fixed-effects estimator.[5] We also explored models in which time dummies were included to control for autonomous changes in the macroeconomic environment. As these were found to be jointly insignificant across all of the models estimated, they were excluded from the final specifications.

Table 3.2 Estimation results for Model 1 and the rebound based on Definition 1

ln(*s*)		*Fixed-effects estimator*			*Between-groups estimator*			*Random-effects estimator*	
		Coeff.s	*Std. Errors*		*Coeff.s*	*Std. Errors*		*Coeff.s*	*Std. Errors*
ln (μ)	**	.585	(.122)	**	.575	(.139)	**	.584	(.105)
ln (p_e)	**	−.622	(.164)	**	−.595	(.211)	**	−.588	(.127)
car age		−.011	(.006)	**	−.02	(.006)	**	−.018	(.005)
diesel		−.232	(.182)		.12	(.102)		.014	(.090)
premium car		.14	(.147)	**	.252	(.062)	**	.22	(.052)
household size		−.013	(.050)		.003	(.031)	*	.04	(.025)
# high school diploma		−.012	(.055)	**	.095	(.036)	*	.07	(.030)
# employed	**	−.135	(.046)	**	.208	(.037)	**	.092	(.029)
car vacation	**	.275	(.037)	**	.411	(.070)	**	.3	(.031)
children		−.064	(.118)		.062	(.091)		.036	(.067)
job change		.113	(.071)		.127	(.089)	**	.145	(.055)
constants		–	–	**	7.768	(.284)	**	7.854	(.219)
$H_0 = \alpha_\mu = -\alpha_{p_e}$		F(1, 545) = 0.03			F(1, 545) = 0.01			X^2 (1) = 0.02	
Hausman Test:					X^2 (11) = 55.19**				

Note: * denotes significance at the 5% level and ** at the 1% level, respectively. Number of observations used for the estimations: 1,357. Number of households: 546. $X^2(11)$ denotes the Chi square test statistic with 11 degrees of freedom, reflecting the comparison of 11 parameter estimates obtained either from the fixed- or random-effects estimation.

Table 3.2 presents estimates corresponding to Definition 1 of the rebound effect, in which fuel efficiency is regressed on the distance driven using fixed-, between-groups, and random-effects estimators. Several features of the results bear highlighting. First, the estimated rebound effects are considerably higher than most estimates reported elsewhere in the literature, and suggest that some 58 per cent of the potential energy savings due to an efficiency improvement is lost to increased driving. Second, these effects are of a strikingly similar magnitude across the three estimators, differing by less than a percentage point.

Finally, we confirm that the impact of efficiency improvements on travelled distance is of the same order as the effect of fuel prices: as reported in the penultimate row of the table, upon testing the null hypothesis H_0: $\alpha_\mu = -\alpha_{p_e}$, we cannot reject the anti-symmetry given by H_0 for any of the estimation techniques. Hence, there is no reason, either on a theoretical or an empirical basis, to assume that Models 1 and 2 are principally different, implying the conclusion that it is equally well-founded to estimate the direct rebound effect on the basis of either Definition (Model) 1 or Definition (Model) 2.

We consequently refrain from presenting the results of Model 2, though we note that the rebound estimates are only negligibly different from those presented in Table 3.2. It is also worth noting that most other studies have not tested this anti-symmetry hypothesis and those few studies that have examined this hypothesis have either been unable to reject it as well or found that the fuel efficiency elasticity is less than the fuel cost per mile elasticity (Sorrell and Dimitropoulos, 2007).

Turning to the other coefficients, we see that the tight correspondence evidenced for the estimate of $\ln(\mu)$ across the models generally does not hold. A particularly stark difference is seen for the effect of the number of employed household members, which has a counterintuitive and negative coefficient in the fixed-effects model, but is positive in the between-groups and random-effects models. All else equal, we would expect that a greater number of employed persons in the household would increase the dependency on the automobile. The negative fixed-effects estimate is difficult to interpret, but may be the result of temporary disruptions associated with changes in the household labour force that reduce automobile travel. Given this difference, it is not surprising that a Hausman test rejects the null hypothesis that the fixed- and random-effects coefficients are jointly equal for all significance levels.

The remaining control variables have either intuitive effects or are statistically insignificant. Referring to the random-effects coefficients, older cars are seen to be driven less, while premium cars are driven more. Another important determinant is whether a vacation with the car was undertaken over the survey period, which results in a roughly 35 per cent ($= \exp(0.30) - 1$) increase in distance travelled. Aside from the fuel price and the number of employed household members, this is the only control variable also found to be significant in the fixed-effects model.

Table 3.3 presents estimates of Model 3, which is distinguished by the use of total fuel consumption as the dependent variable and the price of fuel per litre as the key regressor. That this model is not identical to Models 1 or 2 is confirmed by the rejection of the null hypotheses that the price coefficients in Models 3 and 2 are both -1. Despite these differences, the estimates in Table 3.3 are remarkably similar to those of Table 3.2, albeit with a larger range across the fixed- and between-groups effects estimators. In this instance, the estimated rebound effect is seen to vary between 57 and 67 per cent.

Table 3.3 Estimation results for Model 3 and the rebound based on Definition 3

ln(*s*)		*Fixed-effects estimator*			*Between-groups estimator*			*Random-effects estimator*	
		Coeff.s	*Std. Errors*		*Coeff.s*	*Std. Errors*		*Coeff.s*	*Std. Errors*
$\ln(p_e)$	**	−.569	(.165)	**	−.671	(.211)	**	−.594	(.128)
car age		−.010	(.007)	**	−.019	(.006)	**	−.018	(.005)
diesel car		−.297	(.195)		.001	(.094)		−.076	(.091)
premium car		.149	(.149)	**	.330	(.057)	**	.285	(.050)
household size		.002	(.048)		.027	(.031)	*	.045	(.024)
# high school diploma		.002	(.055)	**	.081	(.036)	*	.065	(.029)
# employed	**	−.117	(.044)	**	.219	(.036)	**	.093	(.029)
car vacation	**	.254	(.036)	**	.412	(.070)	**	.289	(.032)
children		−.020	(.118)		.053	(.092)		.055	(.066)
job change		.116	(.070)		.103	(.089)	**	.137	(.055)
constants		–	–	**	4.003	(.073)	**	4.098	(.063)
H_0: $\alpha_{p_e} = -1$		$F(1,545) = 6.82^{**}$			$F(1,545) = 2.44$			$X^2(1) = 10.09^{**}$	
Hausman Test:					$X^2(11) = 57.57^{**}$				

Note: * denotes significance at the 5% level and ** at the 1% level, respectively. Number of observations used for the estimations: 1,357. Number of households: 546.

We thus conclude that although our estimates of the rebound effect are high, they appear to be robust to both the estimator and the specification. Whether the model controls for time-invariant factors that vary across cases (as with the fixed-effects estimator) or case-invariant factors that vary over time (as with the between-groups effects estimator) has no substantial impact on the key results. Perhaps even more notable is the similarity of the estimates corresponding to Definition 3 with those of Definition 1. While this definition incorporates efficiency via the kilometres per litre travelled, Definition 3 relies exclusively on the price mechanism, suggesting that this information can serve as a useful substitute in the absence of data on technology.

A few caveats should be recognised, perhaps the strongest of which is the assumption that automobile efficiency is exogenous. If, for example, individuals who drive more also select more fuel-efficient vehicles, then we might expect an upward bias imparted on the rebound effect estimated by the coefficients of $\ln(\mu)$ and $\ln(p_s)$. However, there are two reasons why we do not deem endogeneity to be a serious concern here. First, any time-invariant unobservable factors that would otherwise induce correlation between the rebound effect and the error term (e.g. proximity to public transit, environmental attitudes) will be captured by the fixed-effects model. Although we cannot exclude the possibility of relevant time-variant unobservables, we believe the range of included explanatory variables – the presence of young children, job changes, and the number of employed – provides reasonably good coverage of temporal changes whose absence could induce biases. Second, endogeneity problems relating to vehicle choice would not be expected to afflict the estimates from Definition 3 in Table 3.3, as these are based on the price of fuel. The fact that these estimates are of roughly the same magnitude as those from Definitions 1 and 2 provides some confirmation that any upward bias from endogeneity is negligible.

An additional caveat arises from the analysis' neglect of changes in capital costs due to efficiency improvements in automotive technology. Specifically, more costly equipment reduces disposable income, thereby triggering an income effect that mitigates the rebound. Several authors, such as Henley et al. (1988), therefore argue that neglecting capital cost would lead to an overestimation of the rebound when relying on definitions of the direct rebound effect, such as Definitions 1, 2, and 3. Moreover, if higher income groups value their time more, the associated rebound effect for these groups may be lower, since they are less inclined to spend more time on travelling. As the mobility data analysed here include no information on annualised capital cost k_i, nor time requirements, nor the consumption of other goods o_i, we are not able to pursue these issues. To date, only a few empirical studies, such as Brännlund et al. (2007), have employed data that allow for explicit consideration of income effects on the consumption of goods other than fuel, thereby enabling estimation of the total rebound effect, including the indirect effect. This is clearly an area warranting further study.

A final caveat pertains to the exclusion of multiple-car households from the models. To the extent that such households enjoy greater substitution possibilities, they would be better positioned to alter automobile efficiency in response to changes in fuel prices by switching cars. As the relatively small share of multiple-car households in our sample precluded detailed analysis of this issue, our rebound effect estimates are only valid for single-car households. For multiple-car households, we might expect to find a smaller rebound effect given that switching to fuel-saving cars may be a strategy that substitutes for reducing vehicle kilometres.

Summary and conclusion

Industrialised countries are increasingly struggling both to ensure their security of energy supply and to reduce emissions of greenhouse gases. It is commonly asserted that efficiency-increasing technological innovations, particularly in the transport sector, are an important pillar in this process. This assertion underpins the CAFE standards in the United States and the more recently reached voluntary agreement between the European Union and the European Automobile Manufacturers Association (ACEA) stipulating the reduction of average emissions in the new car fleet.

Although increased efficiency confers economic benefits in its own right, its effectiveness in reducing fuel consumption and pollution depends on how consumers alter behaviour in response to cheaper energy services due to improved efficiency. To the extent that service demand increases via rebound effects, gains in reducing environmental impacts and energy dependency will be offset. The results presented in this chapter, based on the analysis of a German household panel, suggest that the size of this offset is potentially quite large, varying between 57 and 67 per cent. Stated alternatively, the relative reduction in energy use due to a percentage change in efficiency is in the order of 33–43 per cent.

While these estimates are considerably higher than those found elsewhere in the literature, with most empirical evidence originating from the US, our results are robust to both alternative panel estimators and to alternative measures of the rebound effect. Moreover, our results are consistent with recent anecdotal evidence from Germany. Between 2004 and 2005, fuel prices increased by 5 per cent while average road mileage decreased by 3 per cent (MWV, 2007), suggesting a sizeable price elasticity of −0.6. One possible explanation for the discrepancy between the estimates in this study and those from the US may be superior access to public transport in Germany. Related to this, longer trip distances in the US, particularly for commuting (Stutzer and Frey, 2007), likely decrease the responsiveness of mode choice to changes in automobile efficiency.

As this is one of the few studies to be conducted on this issue in a European context, it would be of interest to see whether the findings presented here are corroborated by studies using other data sets from within Germany and other European countries. If this is found to be the case, it would suggest that policy interventions targeted at technological efficiency – be they voluntary agreements or command and control measures – may have only muted effects in reducing fuel consumption. At the very least, our results indicate that the current emphasis on efficiency as the principal means for policy-makers to address environmental challenges may be misplaced. Given the strong responses to prices found here, price-based instruments such as fuel taxes would appear to be a more effective policy measure.

Technical annex

Largely following Greene (2003: 283–302), we will present here the technical details pertaining to the three principal estimation approaches typically employed for analysing panel data: the fixed-, between-groups, and random-effects estimators. Departing from a standard panel data model,

$$y_{it} = \beta^T x_{it} + \mu_i + u_{it}, \quad i = 1, \ldots, N, \; t = 1, \ldots, T, \tag{12}$$

where μ_i designates time-invariant individual effects, u_{it} time-varying idiosyncratic disturbances, N the number of units, and T the time horizon, the fixed-effects model assumes that differences across units, such as households, can be treated as unknown parameters μ_i to be estimated. Using a dummy variable D_i for each individual household, respectively, panel model (15) can be estimated using ordinary least squares (OLS):

$$y_{it} = \beta^T x_{it} + \mu_i D_i + u_{it}, \quad i = 1, \ldots, N, \; t = 1, \ldots, T. \tag{13}$$

Panel model (16) is usually referred to as the least squares dummy variable (LSDV) model (Greene 2003: 287).

The between-groups effects estimator is simply based on the OLS regression of group means:

$$\overline{y}_i = \beta^T \overline{x_i} + \mu_i + \overline{u_i}, \quad i = 1, \ldots, N, \tag{14}$$

where $\overline{y_i}\!: = 1/T \sum_t^T y_{it}$ denotes the time average of the dependent variable for an individual household i, with $\overline{x_i}$ and $\overline{u_i}$ being the corresponding averages of the explanatory variables and the residuals. Note that estimating (17) involves only N observations, the group means. Hence, most of the available information is ignored by the between-groups effects estimator, which is only unproblematic if there is no variability across time at all. This is the reason for the fact that empirical panel studies rarely report between-groups effects.

Instead of using the dummy variable approach (16), the fixed effects can also be obtained from estimating the following model:

$$y_{it} - \overline{y_i} = \beta^T(x_{it} - \overline{x_i}) + u_{it} - \overline{u_i}, \quad i = 1, \ldots, N, \ t = 1, \ldots, T, \tag{15}$$

which results from subtracting (17) from (15). Note that by taking differences, one can purge the time-invariant individual effects μ_i. In fact, estimating (16) using OLS is equivalent to the least squares regression of $(y_{it} - \overline{y_i})$, the mean deviation of y_{it}, on $x_{it} - \overline{x_i}$, the demeaned set of regressors, as is suggested by (18).

In the random-effects model, finally, the time-invariant, unobservable effects μ_i are treated as part of the disturbances, $\upsilon_{it} := \mu_i + u_{it}$, thereby assuming that their correlation with the regressors is zero:

$$y_{it} = \beta^T x_{it} + \upsilon_{it}, \quad i = 1, \ldots, N, \ t = 1, \ldots T. \tag{16}$$

It is further assumed that there is no correlation between the individual effects μ_i and the idiosyncratic disturbance u_{it}. As the covariance matrix of the error term is generally non-diagonal, with $Cov(\upsilon_{it}, \upsilon_{is}) = \sigma_\mu^2 \neq 0$ $(s \neq t)$ being the non-diagonal elements, the random-effects model has to be estimated using generalised least squares (GLS) in order to obtain efficient estimates.

Note that only if the individual effects μ_i are strictly uncorrelated with the regressors is it appropriate to model the individual specific constant terms as randomly distributed across households, as is suggested by the random-effects model (19). By contrast, the fixed-effects model (16) explicitly allows the unobserved individual effects μ_i to be correlated with the included variables given by vector x. The cost of this advantage of the fixed-effects model is the large number of parameters to be estimated, thereby substantially reducing the degrees of freedom, and hence efficiency. Conversely, the payoff to the random-effects model is the greater efficiency due to the reduced number of parameters to be estimated. The disadvantage of this efficiency gain, however, is the possibility of getting inconsistent estimates.

Given the distinction between fixed- and random-effects estimators and their different features, a natural question is which estimator should be preferred. The answer to this question is typically based on a Hausman specification test. The underlying null hypothesis states that the fixed- and random-effects coefficients are jointly equal.

$$H_0\colon \beta_{fixed} = \beta_{random}. \tag{17}$$

The null is examined on the basis of the Wald criterion:

$$W = (\widehat{\beta}_{fixed} - \widehat{\beta}_{random})^T (\widehat{\psi})^{-1} (\widehat{\beta}_{fixed} - \widehat{\beta}_{random}), \quad (18)$$

where $\widehat{\psi}$ denotes the covariance matrix of the difference vector:

$$\widehat{\psi} = Var(\widehat{\beta}_{fixed} - \widehat{\beta}_{random}) = Var(\widehat{\beta}_{fixed}) - Var(\widehat{\beta}_{random}). \quad (19)$$

That is, $\widehat{\psi}$ can be obtained from the estimated covariance matrix of the slope estimator in the LSDV model and the estimated covariance matrix in the random-effects model, excluding the constant term. Under the null hypothesis, W has a limiting chi-squared distribution χ^2 with the number of degrees of freedom equalling the number of coefficient comparisons.

Acknowledgements

We are very grateful for valuable comments and suggestions by Steven Sorrell. Furthermore, we would like to thank participants of the Ninth European IAEE Conference 2007 in Florence, Italy, for constructive discussions. This chapter is adapted from a recent paper by the authors (Frondel et al., 2008).

Notes

1. The literature distinguishes between direct and indirect rebound effects (e.g. Greening and Greene, 1997; Greene et al., 1999). The indirect effect arises from a change in income: lower per-unit cost of an energy service implies – *ceteris paribus* – that real income grows. Given that indirect effects are difficult to quantify, the overwhelming majority of empirical studies confine themselves to analysing the direct rebound effect.
2. Of the remaining 47 per cent of German households, 27 per cent have more than one car and 20 per cent have no car (MiD 2007).
3. The price series was deflated using a consumer price index for Germany obtained from Destatis (2007).
4. We also tested for quality-dependent rebound effects by interacting the efficiency measure with the luxury dummy, but found this to be insignificant in all of the specifications.
5. Roughly 18 per cent of households changed cars at least once over the three years of the survey.

References

Baltagi, B. H. (2005) *Econometric Analysis of Panel Data*. Third edition, London: John Wiley & Sons.

Becker, G. S. (1965) 'A theory of the allocation of time', *Economic Journal*, 75: 493–517.
Berkhout, P. H. G., J. C. Muskens and J. W. Velthuisjen (2000) 'Defining the rebound effect', *Energy Policy*, 28: 425–32.
Binswanger, M. (2001) 'Technological progress and sustainable development: what about the rebound effect?', *Ecological Economics*, 36(1): 119–32.
Brännlund, R., T. Ghalwash and J. Nordström (2007) 'Increased energy efficiency and the rebound effect: effects on consumption and emissions', *Energy Economics*, 29: 1–17.
Brookes, L. G. (1990) 'The greenhouse effect: the fallacies in energy efficiency solution', *Energy Policy*, 18(2): 199–201.
Brookes, L. G. (2000) 'Energy efficiency fallacies revisited', *Energy Policy*, 28(6–7): 355–67.
DESTATIS (Federal Statistical Office Germany) (2007). See http://www.destatis.de/
Frondel, M., J. Peters and C. Vance (2008) 'Identifying the rebound: evidence from a German Household Panel', *Energy Journal*, 29(4): 154–63.
Goodwin, P., J. Dargay and M. Hanly (2004) 'Elasticities of road traffic and fuel consumption with respect to price and income: a review', *Transport Reviews*, 24(3): 275–92.
Greene, D. L. (1992) 'Vehicle use and fuel economy: how big is the "rebound" effect?' *Energy Journal*, 13(1): 117–43.
Greene, D. L., J. R. Kahn and R. C. Gibson (1999) 'Fuel economy rebound effect for US household vehicles', *Energy Journal*, 20(3): 1–31.
Greene, W. H. (2003) *Econometric Analysis*. Fifth edition, London: Prentice-Hall.
Greening, L. A. and D. L. Greene (1997) *Energy Use, Technical Efficiency, and the Rebound Effect: a Review of the Literature*, Washington, DC: Report to the Office of Policy Analysis and International Affairs, US Department of Energy.
Greening, L. A., D. L. Greene and C. Difiglio (2000) 'Energy efficiency and consumption – the rebound effect: a survey', *Energy Policy*, 28(6–7): 389–401.
Henley, J., H. Ruderman and M. D. Levine (1988) 'Energy saving resulting from the adoption of more efficient appliances: a follow-up', *Energy Journal*, 9(2): 163–70.
Khazzoom, D. J. (1980) 'Economic implications of mandated efficiency standards for household appliances', *Energy Journal*, 1: 21–40.
Lovins, A. B. (1988) 'Energy saving resulting from the adoption of more efficient appliances: another view', *Energy Journal*, 9(2): 155–62.
MID (Mobilität in Deutschland) (2007) German Federal Ministry of Transport, Building and Urban Affairs. See http://www.kontiv2002.de/publikationen.htm
MOP (German Mobility Panel) (2007). See http://www.ifv.uni-karlsruhe.de/MOP.html
MWV (Association of the German Petroleum Industry) (2007). See http://www.mwv.de/
Saunders, H. D. (1992) 'The Khazzoom-Brookes postulate and neoclassical growth', *Energy Journal*, 13(4): 131–48.
Small, K. A. and K. Van Dender (2007) 'Fuel efficiency and motor vehicle travel: the declining rebound effect', *Energy Journal*, 28(1): 25–52.
Schipper, L. and M. Grubb (2000) 'On the rebound? Feedback between energy intensities and energy uses in IEA countries', *Energy Policy*, 28: 367–88.

Sorrell, S. and J. Dimitropoulos (2007) *Technical Report 2: Econometric Studies, UKERC Review of Evidence for the Rebound Effect*, London: UK Energy Research Centre.

Sorrell, S. and J. Dimitropoulos (2008) 'The rebound effect: microeconomic definitions, limitations, and extensions', *Ecological Economics*, 3: 636–49.

Stutzer, A., and B. S. Frey (2007) 'Commuting and life satisfaction in Germany', *Informationen zur Raumentwicklung*, 2/3: 179–89.

Walker, I. O. and F. Wirl (1993) 'Asymmetric energy demand due to endogenous efficiencies: an empirical investigation of the transport sector', *Energy Journal*, 14: 183–205.

West, S. E. (2004) 'Distributional effects of alternative vehicle pollution control policies', *Journal of Public Economics*, 88(3–4): 735–57.

Wirl, F. (1997) *The Economics of Conservation Programs*, London: Kluwer Academic Publishers.

Part II
Economy-wide Rebound Effects

4
Modelling the Economy-wide Rebound Effect

Grant Allan, Michelle Gilmartin, Peter G. McGregor, J. Kim Swales and Karen Turner

In this chapter we address the question: how large are the system-wide rebound effects likely to be for general improvements in energy efficiency in production activities in a developed economy? Most studies focus on the direct rebound effects, often in the context of the demand for consumer services (Greening et al., 2000; Sorrell, 2007). This restricts the analysis solely to the energy required to provide the consumer services to which the efficiency improvement directly applies. There is, to date, comparatively little evidence on the scale of non-direct or system-wide effects that reflect the relative price, output and income changes induced by improvements in energy efficiency (Greening et al., 2000).

We begin with a brief summary of previous theoretical and empirical analyses. We then use an economy-energy-environment Computable General Equilibrium (CGE) model of the UK to identify the likely system-wide ramifications of a stimulus to industrial energy efficiency.[1]

The general equilibrium rebound effect: theory

It is instructive to provide a brief account of what has come to be known as the economy-wide rebound effect: that is, the economy-wide impact of an improvement in energy efficiency analysed within a general equilibrium framework. We begin by making a distinction between energy measured in natural units, E, and efficiency units, ε. The natural unit measure could be any physical measure of energy, e.g. kWh, BTU or PJ, whilst energy in efficiency units is a measure of the effective energy service delivered.[2] If there is energy augmenting technical progress at a rate ρ, the relationship between the percentage change in physical energy use, $\dot{E}$, and the energy use measured in efficiency units, $\dot{\varepsilon}$, is given as:

$$\dot{\varepsilon} = \rho + \dot{E} \tag{1}$$

Equation (1) implies that for a 5 per cent increase in energy efficiency, a fixed amount of physical energy will be associated with a 5 per cent increase in energy measured in efficiency units. This means that in terms of the outputs associated with the energy use, a 5 per cent increase in energy efficiency has an impact that is identical to a 5 per cent increase in energy inputs, without the efficiency gain.

A central issue in the rebound analysis is the fact that any change in energy efficiency will have a corresponding impact on the price of energy, when that energy is measured in efficiency units. Specifically:

$$\dot{p}_{\varepsilon} = \dot{p}_{E} - \rho \tag{2}$$

where $\dot{p}$ represents proportionate change in price and the subscript identifies energy in either natural or efficiency units. Using the same numerical example, with constant energy prices in natural units, a 5 per cent improvement in energy efficiency generates a 5 per cent reduction in the price of energy in terms of efficiency units.

With physical energy prices constant, we expect a fall in the price of energy in efficiency units to generate an increase in the demand for energy in efficiency units. This is the source of the rebound effect. In a general equilibrium context:

$$\dot{\varepsilon} = -\eta \dot{p}_{\varepsilon} \tag{3}$$

where η is the general equilibrium price elasticity of demand for energy and has been given a positive sign. For an energy efficiency gain that applied across all energy uses within the economy, the change in energy demand in natural units can be found by substituting equations (2) and (3) into equation (1), giving:

$$\dot{E} = (\eta - 1)\rho \tag{4}$$

For an efficiency increase of ρ, rebound, R, is defined as:

$$R = 1 + \frac{\dot{E}}{\rho} \tag{5}$$

Rebound measures the extent to which the change in energy demand fails to fall in line with the increase in energy efficiency. Therefore where rebound is equal to unity, there is no change in energy use as the result of the change in energy efficiency. Rebound values less than unity but greater than zero, reported here as a percentage, imply that there has been some energy saving as a result of the efficiency improvement, but not by the full extent of the efficiency gain. If a 5 per cent increase in energy efficiency generates a 4 per cent reduction in energy use, this corresponds to a 20 per cent rebound.

If equation (4) is substituted into equation (5), the link between rebound and the general equilibrium elasticity of demand for energy is made absolutely clear:

$$R = \eta \tag{6}$$

This conceptual approach is ideal for a fuel that is imported and where the natural price is exogenous or only changes in line with the demand measured in natural units. There are three important ranges of general equilibrium price elasticity values. If the elasticity is zero, the fall in energy use equals the improvement in efficiency and rebound equals zero. If the elasticity lies between zero and unity, so that energy demand is relatively price inelastic, there is a fall in energy use but some rebound effect. Where the elasticity is greater than unity, so that demand is relatively price elastic, energy use increases with an improvement in energy efficiency. The rebound is greater than unity (>100 per cent) and backfire occurs.

However, in an empirical context there are two sources of additional complexity. First, energy is often domestically produced and uses energy as one of its inputs. This means that the price of energy in physical units will be endogenous, giving further impetus for rebound effects. The second is the identification of the general equilibrium elasticity of demand for energy. The responsiveness of energy demand at the aggregate level to changes in (effective and actual) prices will depend on a number of parameters and characteristics of the economy, as our theoretical analysis in Allan et al. (2008) demonstrates. As well as elasticities of substitution in production, which tend to receive most attention in the literature (see Broadstock et al., 2007, for an excellent review) these include: price elasticities of demand for individual commodities; the degree of openness and extent of trade (particularly where energy itself is traded – see Hanley et al., 2006); the elasticity of supply of other inputs/factors; the energy intensity of different activities; and income elasticities of energy demand (the responsiveness of energy demand to changes in household incomes). Thus, the extent of rebound effects is, in practice, always an empirical issue.

Before turning to general equilibrium modelling to investigate the likely scale of the system-wide rebound effects, a final point should be made. In our investigations here we only improve energy efficiency in a subset of its uses: that is, in its use in production. The proportionate change in energy use is therefore $\Delta E_T/E_I$, where the I and T subscripts stand for total and industrial. Similarly, the implied reduction in the price of energy in efficiency units only applies to its use in production, not in elements of final demand, such as household consumption.

The general equilibrium rebound effect: literature

One way of approaching this issue is through simulation, and the most appropriate approach to use is a Computable General Equilibrium (CGE) model. In their survey of the rebound effect, Greening et al. (2000) found only one paper that examined the economy-wide effects of improved efficiency (Kydes, 1997). As Greening et al. (2000: 397) note, 'prices in an economy will undergo numerous, and complex adjustments. Only a general equilibrium analysis can predict the ultimate result of these changes.' Since their review, there has been an expanding (but still small) literature that analyses the system-wide impacts of improvements in energy efficiency using CGE analysis. Sorrell (2007: 41) reports that, unlike the direct rebound effect mentioned above, 'despite the widespread use of these [CGE] models, there are only a handful of applications to the rebound effect'.

A basic introduction to CGE modelling

There is a wide range of general equilibrium model types. They share their roots in Walrasian general equilibrium theory though few CGE models nowadays assume full market clearing or universal perfect competition. The schematic of a typical CGE is reminiscent of the simplest macroeconomic circular flow diagram, and is often characterised by disaggregation of households and industries/commodities. The requirement of general equilibrium is a simultaneous equilibrium in commodity and factor markets at an identifiable set of relative prices. However, within this framework the variation amongst model types reflects a heterogeneity of views with respect to how markets function and the particular model focus (target economy(ies), types of markets and transactors, macroeconomic closures/assumptions, dynamics etc.).

Typically, the core database is a social accounting matrix (SAM), which describes the structure of the economy in terms of flows of income and expenditure between each production and consumption activity, and transfers of income to and from local and external transactors.[3] The parameters of a CGE model are generally determined in three ways. First, structural parameters (e.g. industry cost structures) are given by the base-year data (SAM). Second, key parameters (e.g. substitution elasticities) are identified by econometric estimation and/or informed judgement involving literature review. These parameters are imposed in the model but may be subject to sensitivity analysis. Third, all the remaining parameters are determined through calibration to the base-year data set (the SAM). This involves assuming that the base-year SAM reflects a long-run equilibrium so that running the model with no change in exogenous variables will reproduce this equilibrium.

A fuller non-technical introduction to CGE models is given in Greenaway et al. (1993). Here we focus on a small set of CGE models that simulate the impacts of increased energy efficiency and estimate rebound effects. Eight papers have been identified and in the remainder of this section we outline their assumptions and results, the similarities and differences between them and the mechanisms through which these results are achieved. Such an approach is useful in identifying best practice and possible lessons for future research. Table 4.1 summarises the results and key features of each of the models. While the studies are diverse, it is notable that the lowest estimate for the long-run economy-wide rebound effect is between 14 per cent and 31 per cent (Allan et al., 2006) and that four of the models predict backfire. In the next section we list and discuss the key features of these studies and examine how they differ.

Treatment of energy in the production function

During construction of a CGE model, the developer needs to specify the structure of inputs to production and consumption. In production this typically takes the form of production functions that allow for substitution between different inputs in the production of any given level of output. The parameter that describes the ease of substitution between two inputs is the elasticity of substitution, σ.[4] Many CGE models combine pairs of inputs (e.g. capital and labour) to produce an intermediate, or composite, output that subsequently combines with another input at the next level in the hierarchy. This is termed a nested structure.

Within CGE models there are a range of alternative approaches for combining energy with other inputs, as well a range of choices for the relevant elasticities of substitution. Figure 4.1 illustrates the four different

Table 4.1 CGE studies to date, critical details with each paper and estimated rebound effects

Author/Date	*Country or region*	*Treatment of energy in production function (Figure 4.1)*	*Elasticity of substitution with energy*	*Efficiency improvements*	*Rebound or backfire evident?*
Semboja, 1994	Kenya	Type A between electricity, other fuels, capital and labour.	Cobb-Douglas at this level but Leontief between basic inputs and capital.	Two scenarios: an improvement of energy production efficiency and an improvement in energy use efficiency.	Backfire reported in both scenarios.
Dufournaud et al., 1994	Sudan	Utility functions.	Two values for CES for energy employed in each version of utility functions of 0.2 and 0.4.	Improvement in efficiency which wood-burning stoves can meet household energy from firewood.	Household consumption of energy services increases in all cases, while demand for demand for firewood declines. Rebound estimated as between 47 per cent and 77 per cent.
Vikstrom, 2004	Sweden	Type C in production of value-added composite.	CES at sectoral level. Values range from 0.07 to 0.87.	Single simulation with 15 per cent increase in energy efficiency in non-energy sectors and 12 per cent increase in energy sectors.	Rebound of around 60 per cent reported.
Washida, 2004	Japan	Type B in production of energy and value-added composite.	CES between energy and value added of 0.5.	1 per cent increase in energy efficiency in production and consumption.	53 per cent rebound in central simulation (calculated as Rebound = Improvement rate of efficiency minus Reduction rate of CO_2).

Grepperud and Rasmussen, 2004	Norway	Type C in production of value-added composite.	CES between energy and capital at sectoral level. These differ by sectors, but generally between 0 and 1.	Historically estimated annual average growth rates of energy productivity at the sectoral level are doubled. Four sectors have electricity efficiency doubled, while two have oil efficiency doubled.	Oil efficiency sectors generally exhibit small rebound, while rebound and backfire effects are seen in electricity efficiency improving sectors.
Glomsrød and Taoyuan, 2005	China	Type A in production of value-added composite.	Cobb-Douglas substitution between energy, capital and labour.	Business-as-usual dynamic scenario compared to case where costless investments generate increased investments and productivity in coal cleaning sector, lowering price and increasing supply.	>100 per cent rebound or backfire observed in scenario modelled.
Hanley et al., 2006	Scotland	Type D in production of gross output.	CES between energy and non-energy intermediates of 0.3.	5 per cent improvement in efficiency of energy use across all production sectors (including energy sectors).	>100 per cent rebound or 'backfire' in long run in central simulation.
Allan et al., 2006	UK	Type D in production of gross output.	CES between energy and non-energy intermediates of 0.3.	5 per cent improvement in efficiency of energy use across all production sectors (including energy sectors).	Between 36 and 55 per cent in short run, and 14 and 31 per cent in long run, for electricity and non-electricity energy respectively in each conceptual time period.

types of nested structure that are used within the papers we study. It might be expected that where energy is included in the nested structure will have implications for the model results.

In production structure A in Figure 4.1, used by Semboja (1994) and Glomsrød and Taoyuan (2005), energy is directly substitutable for both capital and labour. In production structure B, energy substitutes with a value-added composite formed by a combination of labour and capital inputs. This treatment is used in Washida (2004). In production structure C, energy and capital combine to produce an energy-capital composite that then substitutes with labour, as used by Vikstrom (2004) and Grepperud and Rasmussen (2004). Production structure D in Figure 4.1 shows the case adopted by Hanley et al. (2006) and Allan et al. (2006), where labour and capital are combined to form a value-added composite and energy and non-energy inputs combine to form an intermediate input composite.

Within a nested production structure, the energy input itself is typically a composite formed by a combination of a number of energy carriers that represent the output of several energy sectors. For instance, the energy composite in Washida (2004) is formed from a combination of oil, coal, gas and electricity inputs, with corresponding assumptions about the ease of substitution between each energy type. The specific method used to construct the energy composite is less important in the present chapter as the improvement in energy efficiency is introduced at the level of the energy composite itself and not on one specific energy carrier.

What can be concluded about the extent to which the location of energy in the nested production structure affects the results? We note that both papers that use A-type functions report backfire. However, without looking at the other key features, we cannot conclude that this result is explained by the nested structure alone. For example, these papers also assume a unitary elasticity of substitution between different inputs, which is greater than suggested by many empirical studies and lower than the values used in the other CGE models. This may play a more important role in producing backfire than the detailed production structure.

Elasticity of substitution with energy in production

There is an acceptance in the rebound literature that the elasticity of substitution of energy for other inputs is important for the size of rebound effects (Broadstock et al., 2007). However, there is less agreement that this

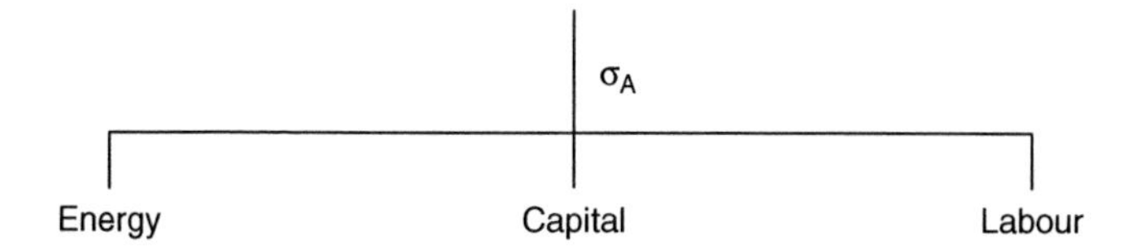

Function B

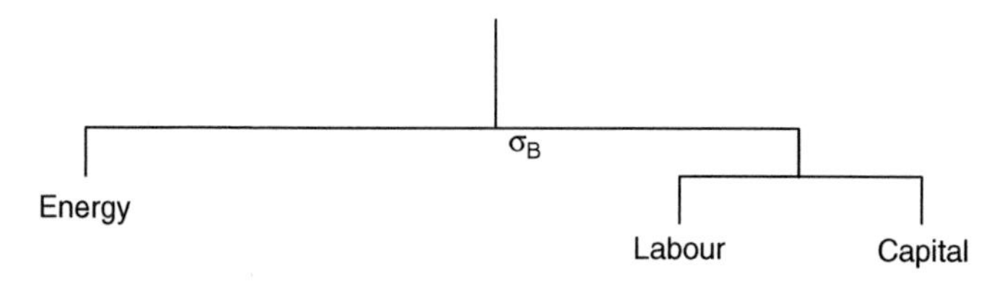

Function C

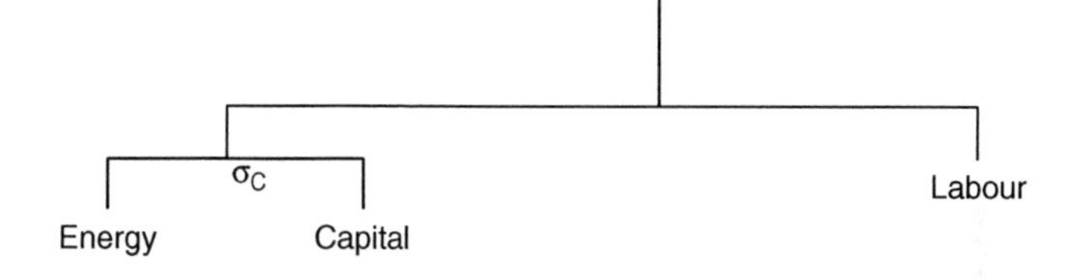

Function D

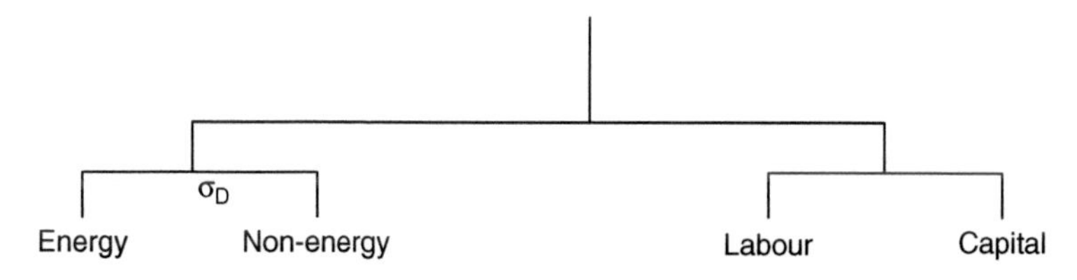

Figure 4.1 Alternative specifications for production functions involving energy

is the single most important elasticity, given the other determinants of rebound effects identified by our theoretical analysis. Here we detail the specific value of the substitution parameter in the production function in each paper (e.g. the value of σ_A, σ_B, σ_C, σ_D parameter in Figure 4.1).

Commonly, the CGE papers reviewed here use a constant elasticity of substitution (CES) specification between energy and other inputs. The nested CES structure used in every paper allows elasticities of substitution to vary between different pairs of inputs. In Washida (2004), where energy substitutes with value added, this elasticity takes a value of 0.5. In Grepperud and Rasmussen (2004) and Vikstrom (2004), where energy substitutes with capital, different values of the elasticity of substitution are used in the production functions for each sector. In Vikstrom (2004) these are taken from contemporary surveys of the relevant literature, and range from 0.07 to 0.87 across sectors. In Hanley et al. (2006) and Allan et al. (2006), energy and non-energy composites substitute with a constant elasticity of substitution of 0.3 for all sectors.[5]

Capital

One crucial component of CGE models is their assumptions about capital stocks. Traditionally, capital stocks are assumed to be fixed in the short run and fully adjustable in the long run. Given their potential importance we note the key assumptions of each paper. A range of treatments is apparent: the total capital stock is fixed across sectors (e.g. Dufournaud et al., 1994); total capital is variable, and new investment is allocated across sectors in fixed shares based on sectoral shares of total capital in the base year (Glomsrød and Taoyuan, 2005); total capital stock is variable through investment, and the sectoral use of capital driven by the sectoral return on capital (e.g. Hanley et al., 2006 and Allan et al., 2006). However, in some papers it is not entirely clear what has been assumed about capital.

The labour market

We know that the specification of the labour market can significantly influence the macroeconomic impact of any disturbance (Harrigan et al., 1996; Gilmartin et al., 2007). It is not surprising that this may also be important for the scale of rebound effects, given the dependence of the latter on induced output changes. Where labour supply is fixed, for example, a positive supply-side policy, such as an increase in the efficiency of energy use, would raise the real wage by stimulating labour demand but would have no impact on employment. If, however, labour supply responds positively to rises in the real wage there will be an expansion in employment that is likely to generate additional rebound effects.

As with the capital market closure, the labour market assumptions are not always made explicit in these papers. In the energy-efficiency simulations reported in Vikstrom (2004) and Washida (2004), it appears that there is a fixed aggregate supply of labour. In contrast, Glomsrød and Taoyuan (2005) assume that there is an infinitely elastic labour supply. In Hanley et al. (2006) wages are determined through a bargained real wage function in which the real consumption wage is directly related to workers' bargaining power and therefore inversely related to the unemployment rate (e.g. Minford et al., 1994). Given the potential importance of the specification of the labour market for estimating rebound effects, there is a disappointing lack of transparency in the models considered here.

Recycling of government revenue

All of the studies model energy-efficiency improvements as a stimulus to the supply side. The two papers that focus on developing countries (Dufournaud et al., 1994; Glomsrød and Taoyuan, 2005) acknowledge the development potential of improvements in energy efficiency as they encourage economic growth and increased employment – and thereby increased taxation revenues. Further, Allan et al. (2006) show that the route through which these revenues are recycled back to the economy can have an important influence on the scale of estimated rebound effects.

Allan et al. (2006) recycle increased government revenue through either increased government expenditure or lower tax rates. In Grepperud and Rasmussen (2004) government expenditure is exogenous and assumed to grow at a constant rate, so that there appears to be no explicit recycling of increased revenue. Hanley et al. (2006) make no adjustment for this either, and neither, we understand, does Semboja (1994). In Glomsrød and Taoyuan (2005) and Vikstrom (2004) the increases in government savings are channelled back into the economy through increased savings driving investment, rather than through increased government expenditures.

The modelling of efficiency improvement

The precise way in which the improvement in energy efficiency is modelled is likely to be important in influencing model results. Generally, this is introduced as a step change in the production efficiency of the energy composite good within the relevant production function (Semboja, 1994; Vikstrom, 2004; Washida, 2004; Allan et al., 2006; Hanley et al., 2006). There is a crucial distinction to be made here between those papers that model a notional across-the-board efficiency improvement in all sectors against those where the stimulus is a precise sector-specific energy-efficiency improvement linked to a particular policy intervention.

Estimated rebound effects

The quantitative results from the papers discussed in this section are shown in the final column of Table 4.1. These range from backfire in a number of cases, to rebound (in non-electricity energy consumption) of as much as 55 per cent in the short run, and 31 per cent in the long run (Allan et al., 2006). As noted above, however, defining the boundary of the energy-efficiency stimulus is important for identifying the precise amount of rebound or backfire in each paper. In some of the papers, it is not clear how the rebound percentage has been calculated, and whether the appropriate boundary has been used to scale the resulting change in energy use and to calculate the size of the rebound effect. Vikstrom (2004) introduces improvements in energy efficiency in industrial sectors, not final consumption, but does not appear to recalibrate the resulting change in energy use to take account of the portion of energy consumed by the industrial sectors in the base year. Where the boundary issue appears to be less of a problem it is sometimes unclear whether the estimated rebound effect has been calculated correctly. Washida (2004) for instance, introduces a 1 per cent improvement in energy efficiency in production and consumption, but rebound is calculated as this initial improvement minus the estimated reductions in the emissions of CO_2, rather than the reduction in energy used.

Summary

This, albeit brief, review of the existing applications of CGE models to the question of energy efficiency and rebound suggests a number of observations. First, there are only a small number of studies. There are significant advantages to be gained from the use of CGE models for the study of rebound effects following energy-efficiency improvements (see Allan et al., 2007a), which makes the limited size of this literature surprising. Second, the studies reviewed here adopt different approaches, making comparison difficult. Third, despite the scope for extensive and informative sensitivity analysis this has not been performed in many of the papers discussed. Such sensitivity might focus on the production structure (i.e. treat the energy composite at different points in the production structure), key parameter values (i.e. varying substitution elasticities at different levels of the production structure) or the labour-market closure. Some elements of this sensitivity are reported for the simulations presented later, although a full systematic sensitivity analysis is beyond the scope of this chapter. Fourth, we note that it is impossible to draw general conclusions with regard to the size of the rebound effect as this will be specific to the economy in question and the particular energy-efficiency stimulus being simulated.

Estimating the UK general equilibrium rebound effect

We now describe our own modelling study of rebound effects in the UK following a 5 per cent improvement in the efficiency of energy use within all production sectors. We first describe the CGE model used (UKENVI), next summarise the simulation outcomes and finally examine the sensitivity results.

The UKENVI model

UKENVI is a CGE model of the UK economy (Allan et al., 2006; Learmonth et al., 2007). It has three transactor groups, namely households, corporations and government; 25 commodities and activities (five of which are energy commodities/activities); and one exogenous external transactor (ROW). In the application described here, commodity markets are assumed to be competitive and we do not explicitly model financial markets.

The UKENVI framework allows a high degree of flexibility in the choice of key parameter values and the treatment of factor and goods markets However, a crucial characteristic of the model is that no matter how it is configured, we impose cost minimisation in production with multi-level production functions in all sectors (see Figure 4.2). There are four major components of final demand: consumption, investment, government expenditure and exports. Of these, in the central case scenario, real government expenditure is taken to be exogenous. Consumption is a linear homogeneous function of real disposable income. Exports (and imports) are generally determined via an Armington link and are therefore sensitive to relative prices (Armington, 1969).[6]

We parameterise the model to be in long-run equilibrium in the base-year period. This implies that the capital stock in each industrial sector is initially fully adjusted to its desired level. The only technical change

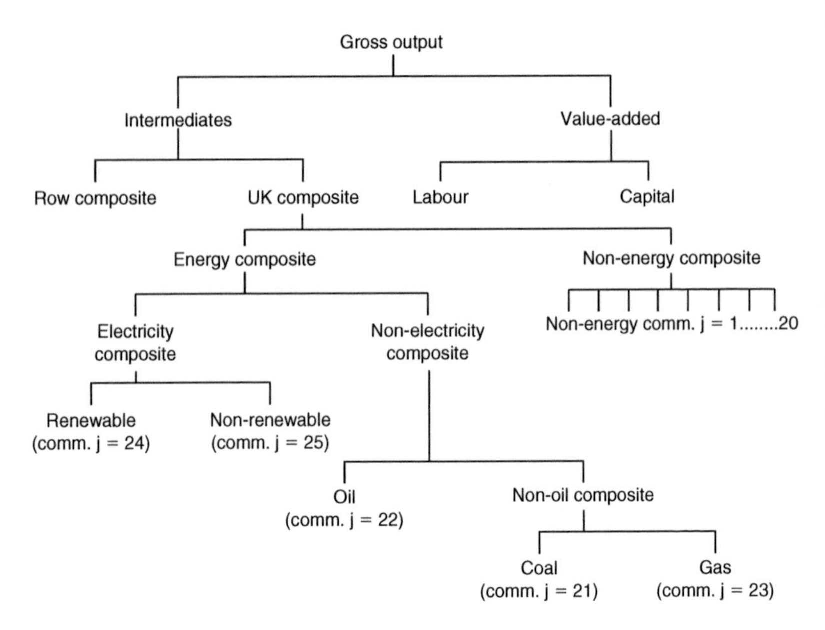

Figure 4.2 Production structure of each sector in the 25 sector/commodity UKENVI framework

introduced in the simulations reported here is the step improvement in energy efficiency. In this chapter we give simulation results for two alternate conceptual time periods: the short run and the long run. In the short run, the capital stock is fixed, both in terms of its absolute size and in its distribution between individual sectors, although labour can move freely between sectors in this time interval. In the long-run, capital stock in each sector adjusts to its new desired level, which depends upon the new values for sectoral value added, the user cost of capital and the wage rate. We assume that interest rates are fixed in international capital markets, so that the user cost of capital varies only with the price of capital goods.

The long run is a conceptual time period. However, where we run the model in a period-by-period mode with the gradual updating of capital stocks, a close adjustment to the long-run values will often take over 25 years. When the model is run in this period-by-period mode, each sector's capital stock is updated between periods via a simple capital stock adjustment procedure, according to which investment equals depreciation plus some fraction of the gap between the desired and actual capital stock.[7] This treatment is wholly consistent with sectoral investment being determined by the relationship between the capital rental rate and the user cost of capital. In the long run, the capital rental rate equals the user cost in each sector, and the risk-adjusted rate of return is equalised between sectors.[8]

We impose a single UK labour market characterised by perfect sectoral mobility. In our central case scenario, wages are determined via a bargained real wage function in which the real consumption wage is directly related to workers' bargaining power, which itself depends inversely on the unemployment rate (Blanchflower and Oswald, 1994; Minford et al., 1994).[9] Here, we parameterise the bargaining function from the econometric work reported by Layard et al. (1991):

$$w_t = \alpha - 0.068u + 0.40w_{t-1} \tag{7}$$

where: w and u are the natural logarithms of the UK real consumption wage and the unemployment rate respectively, t is the time subscript and α is a parameter which is calibrated so as to replicate equilibrium in the base period.[10]

Figure 4.2 summarises the production structure that is imposed in each sector of the model. This separation of different types of energy and non-energy inputs in the intermediates block is in line with the general KLEM (capital-labour-energy-materials) approach that is most commonly adopted in the literature. As discussed later, there is currently no consensus on precisely where in the production structure energy should be introduced. The empirical importance of this choice is an issue that requires more detailed research.

The multi-level production functions in Figure 4.2 are generally of constant elasticity of substitution (CES) form, so there is input substitution in response to relative price changes. In all sectors, the elasticity of substitution at all points in the multi-level production function takes the default value of 0.3 except at two levels of the hierarchy. These are the production of the non-oil and the non-energy composites where, because of the presence of zeros in the base-year data for some inputs within these composites, Leontief (fixed coefficient) functional forms are used.

The main database for UKENVI is a specially constructed Social Accounting Matrix (SAM) for the UK economy for the year 2000. This required the initial construction of an appropriate UK input-output (IO) table since an official UK analytical table has not been published since 1995. A 25-sector SAM was then developed for the UK using the IO table as a major input. Full details on the construction of the UK IO table and

SAM are provided in Allan et al. (2006). The structural characteristics of the UKENVI model are parameterised on the UK SAM.

Simulation results

The disturbance simulated using the UKENVI model is a 5 per cent improvement in the efficiency by which energy inputs are used by all production sectors, including the energy industries themselves. We introduce the energy efficiency shock by increasing the productivity of the energy composite in the production structure of all industries. This procedure operates exactly as in equation (1) and is described as energy augmenting technical change. We do not change the efficiency with which energy is used in final demand: that is, in household consumption, government, investment, tourism or export demand.

The improvement is modelled as a one-off permanent step change in energy efficiency, introduced at the level of the energy composite good (see Figure 4.2). *Ceteris paribus*, this lowers the price of energy, measured in efficiency units, used in production, reducing the price of outputs and stimulating economic activity. Our question is: how will this affect the level of energy consumption in the economy as a whole? In addition, how does the percentage change in total energy consumption compare to the (appropriately weighted) percentage improvement in energy efficiency?

Table 4.2 reports the impacts on key aggregate variables for the central case scenario. This is where the efficiency change is introduced costlessly and the default model configuration and parameter values are used. The figures reported are percentage changes from the base-year values. Because the economy is taken to be in equilibrium prior to the energy-efficiency improvement, the results may be interpreted as being the proportionate changes over and above what would have happened without the efficiency shock.

Results are presented for the short and long run. Recall that in the short run, sectoral capital stocks are fixed at their base-year values, while in the long run, capital stocks adjust fully to their desired sectoral values. With wage determination characterised by a bargained wage curve, a beneficial supply-side policy, such as an improvement in energy efficiency, increases employment, reduces the unemployment rate and increases real wages. This has a positive impact on UK economic activity that is greater in the long run than in the short run. In the long run there is an increase of 0.17 per cent in GDP and 0.21 per cent in employment and exports. The expansion is lower in the short run, where GDP

Table 4.2 Central case results for 5% improvement in industrial energy efficiency (percentage changes from base-year values)

	Short run	*Long run*
GDP	0.11	0.17
Consumption	0.37	0.34
Investment	0.06	0.14
Exports	−0.03	0.21
Imports	−0.23	−0.21
Total employment	0.20	0.21
Real take-home wage	0.28	0.30
Consumer price index	−0.27	−0.27
Total electricity consumption	−2.349	−3.149
Electricity consumption rebound	35.6%	13.6%
Total non-electricity energy consumption	−1.344	−2.050
Non-electricity energy consumption rebound	54.7%	30.9%

increases by 0.11 per cent. Over this time period there is a larger increase in consumption but a fall in exports.

The energy-efficiency improvement primarily increases the competitiveness of sectors that use energy intensively through a reduction in the relative price of their outputs.[11] We begin by considering the long run, where two mechanisms drive this change in competitiveness. First, the increase in energy efficiency raises the overall production efficiency of energy-intensive sectors by the greatest amount. Second, the energy-efficiency improvements in the energy supply industries reduce the relative price of energy, measured in natural units. For both these reasons, energy-intensive sectors experience relatively large reductions in unit costs in the long run, which are passed through to lower prices. The changes in long-run output prices for all production sectors (together with the corresponding short-run changes) for all production sectors are shown in Figure 4.3.[12]

In the long run, real and nominal wages rise as a result of the reduction in the unemployment rate generated by the expansion in economic activity. Nevertheless, the increase in energy efficiency, together with fixed interest rates, is large enough to generate price reductions in all production sectors. However, there are clear sectoral differences that generally reflect variation in the energy intensity across sectors.

The largest price impacts are in the energy sectors themselves, though across these energy sectors there is a non-uniform response. The largest

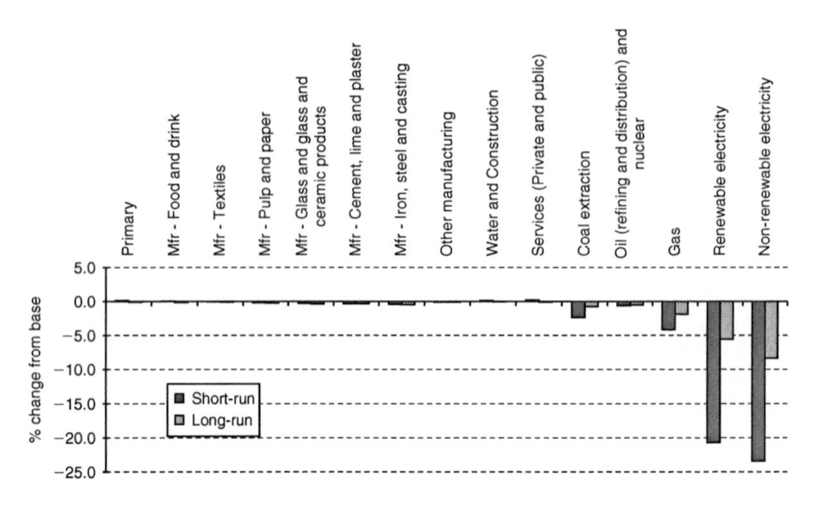

Figure 4.3 The changes in the price of output from a 5% improvement in energy efficiency

reductions in price occur in electricity production, with the price in the non-renewable electricity sector falling by more than the price in the renewable electricity sector. This reflects the heavier reliance of non-renewable electricity production on energy inputs. In the UK SAM for 2000, the renewable and non-renewable electricity sectors purchase 41 and 52 per cent of their inputs respectively from the combined five energy sectors (sectors 21–25 in UKENVI).[13]

Turning now to the short run, in this time interval fixed capital stocks mean that in each sector the marginal cost of the production of value added increases with output. An increase in the demand for a sector's value added in the short run therefore leads, *ceteris paribus*, to an increase in the price of value added, with a corresponding rise in the sector's capital rental rate (see note 8). On the other hand, where the demand for a sector's value added falls, the value-added price will fall, as will the capital rental rate. This form of cost variation is therefore an additional source of changes in relative prices in the short run.

One interesting short-run result is that the output price actually increases slightly in most of the non-energy sectors. These price increases can be traced to a high ratio of value added to gross output in these sectors and the fact that in all these sectors there is a slight increase in both the nominal wage, as reported in Table 4.2, and the capital rental rate. On the other hand, the price adjustments in the energy production sectors are quite different. First, both in the short and long run, energy

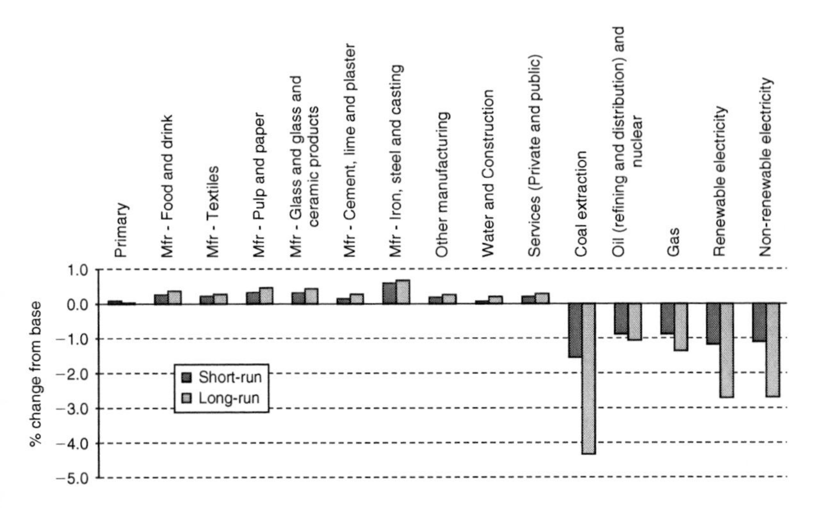

Figure 4.4 The changes in sectoral output from a 5% improvement in energy efficiency

prices fall. Second, the price reduction in the short run is greater than in the long run. Third, these price changes are particularly apparent for the electricity generating sectors because of their own energy-intensities.

In energy sectors, the short-run reductions in demand that accompany the increase in energy efficiency, lead to falls in the capital rental rates, as argued above, such that the price reductions over this time interval are greater than those in the long run. For example, there are significant falls in output price in the renewable and non-renewable electricity generating sectors (down over 20 per cent) in the short run, where the capital rental rates fall by 24 per cent and 26 per cent respectively.

The short- and long-run changes in the output of each sector are shown in Figure 4.4. As would be expected, the increased efficiency of energy inputs has expanded the output of all non-energy sectors, with the increase almost always being greater in the long run than in the short run. Outputs increase most in those non-energy sectors that have greater energy intensities. On the other hand, the output of the five energy sectors falls in both the short and long run, with the reduction greater in the long run. There is greater output responsiveness in the long run since capital stocks can be adjusted to their desired level. In this case the adjustment is downwards through dis-investment. The large reductions in energy prices in the short run go some way to offsetting the demand falls that occur in this time period.

Because the exogenous energy-efficiency improvement applies only to the use of energy as an intermediate input, as discussed earlier the rebound effect is calculated as the increase in total economy-wide energy use divided by the initial energy use in production. The results are presented in Table 4.2. Energy is disaggregated into electricity and non-electricity energy consumption. The short-run rebound effect is 36 per cent for electricity and 55 per cent for non-electrical energy use. These rebound effects are mainly determined through demand for energy as an intermediate input in energy supply and manufacturing. In all cases the overall expansion of the economy generates additional final demand energy use, but this is relatively small. Rebound is much reduced in the long run, with a value for non-electricity consumption of 31 per cent, and 14 per cent for electricity consumption.

Sensitivity analysis

Our central case results presented in Table 4.2 are dependent upon the structural data embedded in the base-year UK SAM. However, they are also sensitive to the choice of key parameters values in the UKENVI model, the recycling of additional government revenues, additional costs that might accompany energy-efficiency improvements and the nature of the labour-market closure. In the next four subsections we outline the effects of varying these assumptions.[14]

Varying key production elasticities

We begin by discussing simulation results where some key parameter values are varied. Figure 4.2 suggests that the production parameters that are expected to affect the results most strongly are the elasticity of substitution between energy and non-energy intermediate composites and between value-added and intermediate inputs in the production of gross output. In the central scenario, both these parameters take the value of 0.3. For sensitivity analysis, we vary these parameters (independently) to 0.1 and 0.7.

Table 4.3 reports the long-run results for a range of simulations. Column A is the central case scenario, so that the figures in this column replicate those in column two of Table 4.2. Columns B and C of Table 4.3 show the long-run results produced by varying the elasticity of substitution between energy and non-energy intermediate inputs.[15] Increasing this elasticity, thereby generating a greater degree of substitutability between energy and non-energy inputs, results in a marginally higher GDP impact. However, as expected, changing this parameter has

Table 4.3 Long-run results for sensitivity analysis for 5% improvement in industrial energy efficiency (% change from base year)

	Central case	*Elasticity of substitution between energy and non-energy composite*		*Elasticity of substitution between value added and intermediate input*		*Impact of recycling government revenues*		*Costly policy simulation*
Simulation reference	*A*	*B*	*C*	*D*	*E*	*F*	*G*	*H*
Parameter values in sensitivity analysis		0.1	0.7	0.1	0.7	Public expenditure adjusts	Income tax adjusts	Increased labour costs
GDP	0.17	0.16	0.18	0.18	0.15	0.20	0.34	−0.33
Consumption	0.34	0.34	0.35	0.35	0.33	0.40	0.69	−0.04
Investment	0.14	0.12	0.17	0.15	0.10	0.13	0.33	−0.32
Exports	0.21	0.21	0.24	0.20	0.26	0.01	0.19	−0.25
Imports	−0.21	−0.21	−0.21	−0.22	−0.20	−0.02	0.03	−0.09
Total employment	0.21	0.21	0.21	0.21	0.20	0.26	0.38	0.03
Real take-home wage	0.30	0.30	0.30	0.31	0.29	0.37	0.55	0.04
Consumer price index	−0.27	−0.23	−0.24	−0.22	−0.25	−0.13	−0.22	0.17
Total electricity consumption	−3.149	−3.720	−1.993	−3.625	−2.181	−3.165	−2.911	−4.132
Electricity consumption rebound	13.6%	No rebound	45.3%	0.5%	40.2%	13.2%	20.1%	No rebound
Total non-electricity energy consumption	−2.050	−2.574	−0.987	−2.314	−1.514	−2.061	−1.788	−3.073
Non-electricity energy consumption rebound	30.9%	13.2%	66.7%	22.0%	49.0%	30.5%	39.7%	No rebound

a major impact on energy use and the extent of rebound. Electricity consumption, which registers no rebound with the low elasticity of substitution (B), does so when the elasticity is increased (C), with a rebound of 45 per cent, and the strength of the rebound effect is substantially increased for non-electricity energy consumption (from 13 per cent to 67 per cent).

In simulation B we record a zero rebound result for electricity consumption. In fact there is a small negative rebound of 2 per cent: a phenomenon that Saunders (2008) refers to as super-conservation. The relevant fall in electricity consumption is proportionally greater than the efficiency improvement. Two points should be made about this result. First, recall that in this simulation an efficiency shock is being applied to both electricity and non-electricity energy use in production. Therefore the change in the electricity use is, in effect, the result of two simultaneous efficiency shocks. Note that we still observe a 13 per cent rebound in non-electricity energy consumption. However, the second point is that where there are changes in the composition of output as a result of efficiency improvements, a very wide range of energy use outcomes are possible. Certainly super-conservation cannot be ruled out for energy efficiency shocks to individual sectors or specific final demand uses. However, a general equilibrium approach is required to capture such changes.

Columns D and E in Table 4.3 give the long-run results when the elasticity of substitution between value added and intermediate inputs in the production of gross output is varied. This is the point at the top level of the production hierarchy shown in Figure 4.2. Again, increasing the ease with which sectors can substitute towards the now cheaper intermediate inputs, including energy, limits the fall in energy use and therefore increases rebound.

Government revenue recycling

In the central case, and in the sensitivity simulations performed up to now, the improvement in energy efficiency stimulates aggregate economic output and employment. In these simulations, with fixed tax and benefit rates, the tax revenues of the UK government rise and social security spending falls. In our central case, the government saves all of this improvement to its budgetary position. We now investigate the implications of recycling these budgetary savings.

In the UKENVI model, the recycling of revenue is achieved by imposing a government budget constraint. This holds constant the base-year ratio of the budget deficit to GDP. However, the government budget constraint can be imposed in two ways that have quite different economic impacts. First, additional revenue can be used to expand general government expenditure, distributed across sectors using the base-year weights. Second, extra revenues received can be returned to households through reduced income tax. The long-run impacts of these two scenarios are shown in Columns F and G of Table 4.3.

Recycling the additional tax revenues as extra government expenditure acts as an exogenous demand injection, which stimulates GDP and employment with changes in output focused on government sectors. Recycling the increased revenues through a reduction in income tax rate produces a more complex set of effects. On the demand side, lowering the tax rate raises take-home wages and thus stimulates household consumption demand. But given a bargained wage curve, there is also a supply-side impact. A fall in the income tax rate means that a lower nominal wage is needed to maintain a given real take-home wage. At any given level of unemployment, nominal wages will fall, stimulating employment. This combination of demand- and supply-side influences means that the recycling through tax reduction generates a significantly larger impact than increased government expenditures.

The energy impacts are also rather different under these two alternative forms of recycling. When government expenditure adjusts, total electricity consumption actually falls by more than under the central scenario. This reflects the switch towards less energy-intensive government demands, and increased employment in the public sector crowding out more energy-intensive private sector activity. When income tax rates adjust, however, electricity use falls by less and rebound increases. For example, for non-electricity energy production, rebound increases from 31 per cent in the central case simulation to 40 per cent where increased revenues are recycled through lower taxes.

Implementing a costly energy efficiency policy

Our assumption up to this point has been that energy-efficiency improvements are delivered costlessly, like manna from heaven. It is useful to consider the case where the gains in energy efficiency are accompanied by other costs borne by production sectors. In the simulations presented here these costs are introduced as an increase in labour costs generated through a reduction in the labour efficiency parameter. This reduction in labour efficiency can be thought of as representing additional labour required to implement the improvement in energy efficiency.[16] The size of the labour efficiency loss introduced in each sector is just enough to counter the impact of the increase in energy efficiency on overall production costs.

The aggregate results from simultaneously introducing both the positive energy-efficiency improvement and the negative shock to the productivity of labour inputs, are shown in the final column (Column H) of Table 4.3. The costly energy-efficiency simulation generates very different changes in energy use. In the long run, energy consumption falls by more than in the base case. Introducing fully-offsetting labour costs related to energy-efficiency improvements neutralises the overall cost reduction and thus inhibits any significant demand switch towards the output of more energy-intensive sectors, so that no rebound effects are identified in this case.

Varying the labour-market closure

In the simulations reported in Table 4.3 the labour market is characterised by a bargained real wage function in which the real consumption wage is inversely related to the unemployment rate. In this sensitivity analysis, two alternative specifications of the labour market are considered. In one we impose an exogenous and inelastic labour-supply function, in which aggregate employment is effectively fixed. This is quite a common assumption in national CGE models, but seems unduly restrictive. In the other specification we use a closure where the real wage is maintained exogenously at the base-year level and employment adjusts to ensure equilibrium. This simulation could also be thought to emulate an economy with no restrictions on immigration (McGregor et al., 1996).[17] These closures can be considered as limiting cases reflecting zero and infinite labour-supply elasticities with respect to the real consumption wage.

Table 4.4 shows the aggregate results in the short and long run from introducing a 5 per cent energy-efficiency improvement in production under these two alternative labour-market closures. In the exogenous labour-supply scenario there is, as would be anticipated, a significant increase in the real take-home wages, but also a small decrease in GDP in each time interval. The GDP change is the result of induced changes in the sectoral composition of output. The fall in energy use is greater than in the central case in this scenario, so rebound is reduced. However, in the real wage resistance scenario, there is a bigger energy rebound than under the central case. For electricity use, for example, the short- and long-run total rebound figures of 36 per cent and 14 per cent are increased to 39 per cent and 34 per cent respectively. Similar adjustments occur with non-electrical energy use.

Table 4.4 Short- and long-run results for 5% improvement in industrial energy efficiency for different labour-market closures (% change from base year)

Labour-market closure	*Exogenous labour supply*		*Real wage resistance*	
Time period	*Short run*	*Long run*	*Short run*	*Long run*
GDP	−0.02	−0.04	0.23	0.90
Consumption	0.30	0.21	0.43	0.80
Investment	−0.15	−0.07	0.27	0.86
Exports	−0.19	−0.03	0.13	1.12
Imports	−0.24	−0.17	−0.21	−0.34
Total employment	0.00	0.00	0.38	0.95
Real take-home wage	0.57	0.39	0.00	0.00
Consumer price index	−0.20	−0.10	−0.39	−0.68
Total electricity consumption	−2.468	−3.357	−2.234	−2.413
Electricity consumption rebound effect	32.3%	7.9%	38.7%	33.8%
Total non-electricity energy consumption	−1.460	−2.256	−1.232	−1.321
Non-electricity energy consumption rebound	50.8%	24.0%	58.5%	55.5%

Conclusions

The simulations reported in this chapter suggest that for the UK, we expect a general, across-the-board, improvement in efficiency in energy use in production to have significant rebound effects, but not backfire. The estimated short-run rebound effects are 36 per cent for electricity and 55 per cent for non-electricity energy consumption, whilst the long-run values are 14 per cent for electricity and 31 per cent for non-electricity energy consumption. Increases in energy efficiency will reduce energy use, but not by the full proportionate amount. Moreover, these impacts will vary across energy types. Most of the rebound effect is captured within the demand for energy as an intermediate input. Additional increases in the demand for energy are driven by the accompanying expansion in final demand but these are relatively small.

We employ sensitivity analysis to explore the robustness of our results to changes in key parameter values and model specification. This analysis backs up the generally held view that the sizes of rebound effects are heavily dependent on the substitution possibilities in production in general, and the scope for substitution between energy and non-energy inputs in particular. However, the size of the rebound effect is also sensitive to assumptions made about the nature of the labour market, the time period under consideration and the manner in which additional government revenues are recycled. In one of the sensitivity scenarios where parameter values are varied, we observe super-conservation and this phenomenon should be explored more thoroughly in a general equilibrium context.

Our analysis shows that it is quantitatively important how we treat any changes in government revenues that result from changes in economic activity and employment that accompany improvements in energy efficiency. We have enforced a government budget constraint in which the additional government revenue is recycled through two alternative channels. Recycling this revenue through raising government expenditure delivers a significantly smaller economic impact than recycling it through reducing the average rate of income tax. With the income tax adjustment, the demand side is affected by increased household consumption, while the supply-side is simultaneously stimulated via a lower nominal wage being required to generate any given real take-home wage. This makes substitution towards labour more attractive, and production more competitive, boosting employment and GDP, and increasing rebound to 40 per cent.

The work in this chapter could be usefully extended in a number of ways. First, we have dealt here solely with improvements in energy efficiency across production sectors, and have excluded efficiency gains in transport and household energy use, where a significant portion of energy policy is directed. Second, analysis could be focused on specific energy-efficiency improvements in individual sectors, rather than general, across-the-board improvements applying to all production sectors. As noted earlier, however, we would not rule out finding significant rebound effects when energy-efficiency improvements are restricted to individual sectors, even those that are not themselves intensive users of energy. Third, further exploration should be made of the impacts of introducing costly improvements in technology. Finally, we could explore the possible role of barriers to the adoption of more efficient technologies (e.g. Sorrell et al., 2004).

Acknowledgements

The authors acknowledge support from the UK Energy Research Centre, although this research draws liberally on related research funded by the Engineering and Physical Science Research Council through the SuperGen Marine Energy Research Consortium (Grant references: GR/S26958/01 and EP/E040136/1) and by the Economic and Social Research Council through the First Grants Initiative (Grant reference RES-061-25-0010). The authors are grateful to Horace Herring (Open University), John Dimitropolous (Sussex Energy Group, University of Sussex), Tina Dallman (Department for Environment, Food and Rural Affairs), Allistair Rennie (Department for Environment, Food and Rural Affairs), Ewa Kmietowicz (Department for Business, Enterprise and Regulatory Reform) and especially Steve Sorrell (Science and Technology Policy Research Unit, University of Sussex) for discussion and comments on this and related work.

Notes

1. This chapter draws from material published in Allan et al. (2007a) for the UK Energy Research Centre review on the rebound effect (Sorrell, 2007) and on simulation results also reported in Allan et al. (2007b).
2. This is very similar to the notion of the effective work that an energy supply can deliver as used in the energy literature.
3. A SAM is typically built around an input-output (IO) table that identifies the structure of the economy in terms of flows of goods and services and generation of GDP. IO tables are commonly produced as part of the standard national accounting framework for industrialised countries under the UN System of National Accounts, 1993.
4. We elaborate more on where this parameter is used within the CGE models here and the values that this parameter could take later in the chapter.
5. As Broadstock et al. (2007) note, the elasticities of substitution employed in CGE analyses often bear a rather complex relationship to those estimated in applied econometric studies.
6. The Armington link is essentially a CES function that allows for less than perfect substitutability between goods produced in domestic and non-domestic economies.
7. This process of capital accumulation is compatible with a simple theory of optimal investment behaviour given a simplifying assumption of quadratic adjustment costs.
8. The capital rental rate equals the profit per physical unit of capital.
9. The real consumption wage is the nominal wage deflated by the consumer price index. This contrasts with the real product wage, where the deflator is the price of domestically produced goods.
10. This calibrated parameter plays no part in determining the sensitivity of the endogenous variables to exogenous disturbances, but the initial assumption of equilibrium is an important assumption.
11. Any simultaneity caused by the increased efficiency of energy use lowering costs of energy supply, thereby lowering the price of energy, which is itself then an input to energy supply, is not unusual in CGE models where these effects can be reconciled in the restoration of an equilibrium.

12. In Figures 4.3 and 4.4, for presentational purposes, the degree of sectoral disaggregation within the non-energy sectors has been reduced.
13. The high value for energy purchases by the electricity sectors reflects the disaggregation into renewable and non-renewable generation of the initial electricity sector in the UK IO table. In the UK electricity sector there are large intra-sector purchases of electricity, which reflect the purchases of electricity from electricity generation sectors to the electricity distribution sector. There are no official data that permit disaggregation of this electricity sector into all activities covering generation, transmission, distribution and supply. Disaggregation of the production structure to distinguish between electricity generation and non-generation is relatively rare in CGE models, although some recent examples do exist (e.g. Wissema and Dellink, 2007). Recent input-output work (Allan et al., 2007c) has disaggregated electricity generation from non-generation activities and further disaggregated generation by technology. This work confirms that energy intensities differ significantly across generation technologies.
14. Research is also required into the precise system-wide rebound effects resulting from sector-specific improvements in energy efficiency. Even with individual sectors being the target of such improvements, there will be a system-wide rebound effect that will be driven by what happens in other sectors, including those not directly affected by the beneficial efficiency improvement. As such, indirect energy intensities, as well as direct energy intensities, will be important. The precise scale of the rebound effect will be related not only to the nature of the individual sector, but also to the economy under consideration.
15. It is useful to look at the empirical evidence concerning the substitutability of energy inputs. Howarth (1997) refers to research that concludes that the elasticity of substitution between energy and non-energy inputs is less than unity. This included a reference to the work of Manne and Richels (1992) who estimated an elasticity of substitution of 0.4 between energy and value-added. The Greening et al. (2000) extensive survey of US work reports some studies that have found an elasticity of substitution greater than one (Hazilla and Kop, 1986; Chang, 1994), but the vast majority of estimates are less than unity. The range of substitution elasticities found in the empirical literature is therefore broadly consistent with the range over which we simulate here. See also note 5.
16. For instance, this could take the form of employing somebody to enforce the energy-efficient technology. However, in principle the same sort of effect can be introduced by reducing the capital efficiency parameter, thereby imposing increased capital costs. More details on this can be found in Allan et al. (2007b).
17. In the central case simulations we assume population is fixed.

References

Allan, G. J., N. D. Hanley, P. G. McGregor, J. K. Swales and K. R. Turner (2006) *The Macroeconomic Rebound Effect and the UK Economy. Final Report to DEFRA*, May. Online at http://www.defra.gov.uk/science/project_data/DocumentLibrary/EE01015/EE01015_3553_FRP.pdf

Allan, G. J., M. Gilmartin, N. D. Hanley, P. G. McGregor, J. K. Swales and K. R. Turner (2007a) *Assessment of Evidence for Rebound Effects from Computable General Equilibrium Modelling Studies. Report for UKERC Technology and Policy Assessment on the Rebound Effect*, London: UK Energy Research Centre.

Allan, G. J., N. D. Hanley, P. G. McGregor, J. K. Swales and K. R. Turner (2007b) 'The impact of increased efficiency in the industrial use of energy: a computable general equilibrium analysis for the United Kingdom', *Energy Economics*, 29: 779–98.

Allan, G. J., N. D. Hanley, P. G. McGregor, J. K. Swales and K. R. Turner (2007c) 'Impact of alternative generation technologies on the Scottish economy: an illustrative input-output analysis', *Proceedings of the Institute of Mechanical Engineering: Part A, Journal of Power and Energy*, 221: 243–54.

Allan, G. J., M. Gilmartin, P. G. McGregor, J. K. Swales and K. R. Turner (2008) 'Energy efficiency', in L. Hunt (ed.), *International Handbook on Economics of Energy*, Cheltenham: Edward Elgar.

Armington, P. (1969) 'A theory of demand for products distinguished by place of production', *IMF Staff Papers*, 16: 157–78.

Blanchflower, D. G. and A. J. Oswald (1994) *The Wage Curve*, Cambridge, MA: MIT Press.

Broadstock, D. C., L. Hunt and S. Sorrell (2007) 'Technical Report 2: Elasticity of Substitution Studies', working paper for UKERC Review of Evidence for the Rebound Effect, ref: UKERC/WP/TPA/2007/011, October.

Chang, K. P. (1994) 'Capital-energy substitution and the multi-level CES production function', *Energy Economics*, 16: 22–6.

Dufournaud, C. M., J. T. Quinn and J. J. Harrington (1994) 'An applied general equilibrium (AGE) analysis of a policy designed to reduce the household consumption of wood in the Sudan', *Resource and Energy Economics*, 16: 69–90.

Gilmartin, M., D. Learmonth, P. G. McGregor, J. K. Swales and K. Turner (2007) 'The national impact of regional policy: demand-side policy simulation with labour-market constraints in a two-region computable general equilibrium model', *Strathclyde Discussion Papers in Economics*, No. 07-04, July.

Glomsrød, S. and W. Taojuan (2005) 'Coal cleaning: a viable strategy for reduced carbon emissions and improved environment in China?', *Energy Policy*, 33: 525–42.

Greenaway, D., S. J. Leybourne, G. V. Reed and J. Whalley (1993) *Applied General Equilibrium Modelling: Applications, Limitations and Future Developments*, London: HMSO.

Greening, L. A., D. L. Greene and C. Difiglio (2000) 'Energy efficiency and consumption – the rebound effect: a survey', *Energy Policy*, 28: 389–401.

Grepperud, S. and I. Rasmussen (2004) 'A general equilibrium assessment of rebound effects', *Energy Economics*, 26: 261–82.

Hanley, N. D., P. G. McGregor, J. K. Swales and K. R. Turner (2006) 'The impact of a stimulus to energy efficiency on the economy and the environment: a regional computable general equilibrium analysis', *Renewable Energy*, 31: 161–71.

Harrigan, F., P. G. McGregor and J. K. Swales (1996) 'The system-wide impact on the recipient region of a regional labour subsidy', *Oxford Economic Papers*, 48: 105–33.

Hazilla, M and R. J. Kop (1986) 'Testing for separable function structure using temporary equilibrium models', *Journal of Econometrics*, 33: 119–41.

Howarth, R. B. (1997) 'Energy efficiency and economic growth', *Contemporary Economic Policy*, 15: 1–9.

Kydes, A. S. (1997) 'Sensitivity of energy intensity in US energy markets to technological change and adoption', in *Issues in Midterm Analysis and Forecasting*, DOE/EIA-060797, US Department of Energy, Washington, DC: 1–42.

Layard, R., S. Nickell and R. Jackman (1991) *Unemployment: Macroeconomic Performance and the Labour Market*, Oxford: Oxford University Press.

Learmonth, D., P. G. McGregor, J. K. Swales, K. Turner and Y. P. Yin (2007) 'The importance of the local and regional dimension of sustainable development: a computable general equilibrium analysis', *Economic Modelling*, 24(1): 15–41.

Manne, A. S. and R. G. Richels (1992) *Buying Greenhouse Insurance: the Economics of CO_2 Emissions Limits*, Cambridge, MA: MIT Press.

McGregor, P. G., J. K. Swales and Y. P. Yin (1996) 'A long-run interpretation of regional input-output analyses', *Journal of Regional Science*, 36: 479–501.

Minford, P., P. Stoney, J. Riley and B. Webb (1994) 'An econometric model of Merseyside: validation and policy simulations', *Regional Studies*, 28: 563–75.

Saunders, H. (2008) 'Fuel conserving (and using) production functions', *Energy Policy*, forthcoming.

Semboja, H. H. H. (1994) 'The effects of an increase in energy efficiency on the Kenyan economy', *Energy Policy*, 22(3): 217–25.

Sorrell, S. (2007) *The Rebound Effect: an Assessment of the Evidence for Economy-wide Energy Savings from Improved Energy Efficiency*, a report published by the Sussex Energy Group for the Technology and Policy Assessment function of the UKERC Energy Research Centre, October.

Sorrell, S., E. O'Malley, J. Schleich and S. Scott, S. (2004) *The Economics of Energy Efficiency: Barriers to Cost-effective Investment*, Cheltenham: Edward Elgar.

Vikstrom, P. (2004) 'Energy efficiency and energy demand: a historical CGE investigation on the rebound effect in the Swedish economy 1957', paper presented at Input-Output and General Equilibrium Data, Modelling and Policy Analysis, Brussels, 2–4 September.

Washida, T. (2004) 'Economy-wide model of rebound effect for environmental policy', paper presented at International Workshop on Sustainable Consumption, University of Leeds, 5–6 March; also presented at ERE W3 conference, Kyoto, Japan, July 2006.

Wissema, W. and R. Dellink (2007) 'AGE analysis of the impact of a carbon energy tax on the Irish economy', *Ecological Economics*, 61: 671–83.

5

Specifying Technology for Analysing Rebound

Harry D. Saunders

Especially in the industrial and commercial sectors, economists seeking to understand rebound phenomena rely on the specification of what they call production functions. As part of this endeavour, they need to mathematically depict the technology gains that may lead to rebound. In this chapter, we attempt to bridge the gap between an engineering understanding of what these technology gains must look like in concrete terms and what they must look like to be implementable in an economic representation.

We begin with a mathematical expression showing, in the simplest possible way, how economists need to think about this problem. Once we have this we can switch to the language of the engineer. The mathematical expression is this:

$$Y = f(K, L, E, M, \ldots)$$

In the language of the economist this is a 'production function'. It depicts the relationship between real output Y (this can be anything from overall economic activity to the quantity of a product produced by a particular industrial or commercial process) and the inputs required to produce it (economists call these inputs 'factors' for reasons known only to themselves). We show here the inputs capital (K), labour (L), energy (E), and materials (M). The ellipsis indicates that there will, in general, be others. Pretend for the moment we somehow know what this function looks like.

Existing technology

Now suppose you are an engineer charged with choosing the best option for providing space heating and cooling in a new commercial building.

You must choose, of course, from existing available technology. The owners of this building would like you to minimise the overall cost of space heating and cooling, subject to certain standards of comfort and safety. A particular decision you must make is the trade-off between the capacity of the heating/cooling system and the amount of insulation to be installed in the structure. There may be similar trade-offs having to do with the choice of windows, and so on. To illustrate the economic concepts, let us say you are not constrained by building codes that limit your choices regarding insulation.

We can describe this situation using our production function. The building owners plan to create a product, or output, which we can think of as a particular quantity of commercial space to be offered for lease (a certain number of square metres). You are in charge of making sure this space is comfortable, climate-wise, so we can define more precisely the output Y offered by the owners to the market as a certain quantity/area of comfortable, climate-controlled commercial space.

The insulation required can be considered part of the capital (K) in the above equation. It will be fixed and in place for the life of the structure. To keep things simple (this will not hamper us in getting where we are heading with this), let us for the moment ignore the fact that other energy will be needed for the building (power, for instance) and imagine that fuel for heating/cooling is the only energy needed. The amount of energy (E) used for ongoing provision of heating and cooling will be determined by among other things boiler capacity, or more generally, the heating/cooling capacity of the system you will choose.

This trade-off between insulation and heating/cooling capacity is an economic trade-off between capital and energy. You will position the particular trade-off selected at a point intended to minimise cost. You will select a certain level of insulation and a certain delivery system. This means you will have chosen a certain capital/energy combination. In your choice, you will have considered the cost of the insulation and the cost of the fuel required to feed the system. To the extent you can, you will consider the expected future costs of fuel. A subtlety is that you will also include the capital cost of the energy conversion system (e.g. the boiler), but you can just think of that as part of the capital/energy trade-off.

But if the cost of insulation and fuel had been different, it is likely you would have chosen a different combination. The production function f needs to accommodate this possibility, and indeed all possible combinations you could have chosen. These combinations will be limited

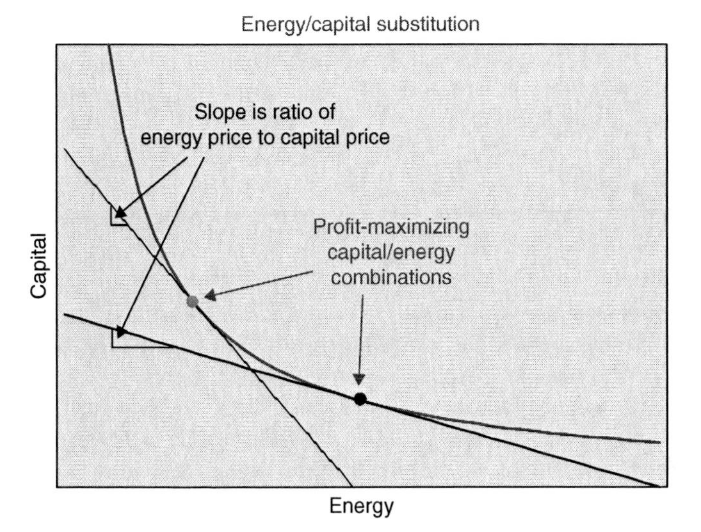

Figure 5.1 Capital-energy substitution

by the technology available to you at the time of your decision – some array of heating/cooling systems and possibly different insulation types. The production function f needs to accommodate this range of possibilities because the economist needs to query this function as to what would happen in the event that prices of fuel and insulation change. The reason the economist needs to know this will become clearer when we consider not existing but new technologies.

But for the moment consider what happens if these prices change. Another engineer, charged with the same task you have undertaken, will perhaps make a different choice of insulation/conversion system (and thus a different capital/energy combination). To an economist, this represents a 'substitution' between capital and energy – or vice versa. A simple diagram is often used to illustrate this (Figure 5.1).

In Figure 5.1, the curved line represents various possible combinations of capital and energy input needed to produce a fixed quantity of Y. You can think of the capital axis as representing different amounts of insulation, and the energy axis as representing different capacities (and therefore fuel usages) of your heating/cooling system. (As noted before, the capital will include the capital cost both of the insulation and of the system itself, but there will be a trade-off looking much like that of Figure 5.1 between total capital and energy capacity.) As you can see, for a very low energy capacity, the amount of insulation needed will be disproportionately large, and for very low insulation energy capacity will be disproportionately large (this disproportionate behaviour is usually the case with input trade-offs of this nature). The cost-minimising combination of capital and fuel will be determined by the relative prices of capital and fuel.

Figure 5.1 shows two different combinations. The black dot represents a cost-minimising combination when fuel price is relatively low and the grey dot represents a cost-minimising combination when fuel price is relatively high. You see that the higher fuel price is associated with lower ongoing fuel use and vice versa. More specifically, it happens that the slopes of the straight lines going through these optimal points are the respective energy price/capital price ratios when costs are minimised. The shift between the two capital/energy combinations, in other words, represents a 'substitution' between capital and energy in response to a change in their relative prices.

A suitable production function f will be able to depict the range of existing substitution possibilities. These possibilities are not always as simple as in our example; the function needs in fact to depict the range of substitution possibilities among the other inputs such as labour and materials. Even in our example, we could imagine that a particular system you choose might be a plant requiring fewer operators than alternative systems. Or perhaps the plant has greater or lesser need of consumable materials, such as lubricants and filters. And so on.

Economists try to develop functions that depict this complex array of possibilities, usually at an aggregate level. They place certain common-sense conditions on what such a function must look like, such as that if you put more input into it the function should produce more output. And that as you try to favour one input over others the harder and harder it should become to produce gains in output. There are other such conditions, but it is possible to produce many functions that behave according to all the common-sense requirements of the economist.

With such a function in hand, economists pass the task to econometricians, who try to measure from real data the parameters of the function. A measured function allows the economist to make certain predictions. So far we have only considered existing technology, but one prediction that in principle can be made with a measured function is what happens to energy consumption when new technology is introduced, and what happens can be subtle and counterintuitive as we will see. So we turn to the topic of new technology.

New technology

Depicting technology at an aggregate level requires understanding it at the application level. Care must be taken to ensure the aggregate depiction conforms to what is really happening in actual applications. This is where the engineer re-enters the picture.

Suppose now you have been charged with the same task, but now you have before you a new alternative. For illustration, let us say this new alternative is a gas-fired (absorption) heat pump, which has only now been invented and was not available to you before. This heat pump can deliver the same amount of space heating/cooling as the system you chose before, but using a fraction of the fuel.

There is a simple and intuitive way to include this new alternative in the production function. This method is not entirely general, but it produces a way to attack the economics problem that is practical and quite rich in what it can deliver by way of insight. Further, it is, we offer, highly consonant with the thinking and language of the engineer. The method is summarised in Box 5.1.

Box 5.1 Generalised technology in production functions

It gives us a clearer picture of what economists have to do to translate engineering efficiency gains into economic terms if we look first at the most general way to do this. Supposing first we have some production function depicting the relationship between real economic output Y and the inputs required to produce it:

$$Y = f(K, L, E, M, \ldots, \boldsymbol{\beta})$$

This simply says that real economic output will depend on the capital, labour, energy, materials, and so on that go into producing that output, and these input 'factors' will deliver this output in a way characterised by the function f and a certain set of parameters $\boldsymbol{\beta}$ (the bold notation just indicates it is a vector, or collection, of several individual parameters).

Then, we ask what happens when we introduce new engineering efficiency gains (new technology, in the parlance of economists). In its most general form, the above expression then becomes

$$Y = f[K(\tau), L(\tau), E(\tau), M(\tau), \ldots, \boldsymbol{\beta}(\tau), \tau(\textit{prices}, \textit{policies} \ldots)]$$

This intimidating-looking expression is actually quite simple. It says there is a collection of technology parameters (the vector $\boldsymbol{\tau}$) that can affect the quantities of inputs or 'factors' used to produce a given level of output. Producers in the economy, upon obtaining new technology, will seek to maximise profit and in so doing may alter the quantities input to production, limited by the production possibilities reflected in the function f and its parameters. The expression also says that this technology may affect the parameters $\boldsymbol{\beta}$ themselves. It further says the magnitude of these technology parameters could in turn depend on the magnitudes of the prices these inputs carry as well as other influencers.

To translate all this into more understandable language, one could say the technology parameters $\boldsymbol{\tau}$ depict how technology could make each input more effective in producing output – in other words, increase the efficiency with which it is used. The dependency of this set of technology parameters on prices and other things, $\boldsymbol{\tau}$(*prices, policies*...), is meant to depict the idea that technology can be endogenously induced – that is to say, can arise in response to, say, increased profit incentives following from changes in prices of the different inputs, or from government incentives or regulations. The dependency of the parameters $\boldsymbol{\beta}$ on the technology parameters, $\boldsymbol{\beta}(\boldsymbol{\tau})$, is meant to indicate that the 'shape' of the production function f could change as a result of new technology arising ('shape' meaning the way this function depicts the space of existing production possibilities).

At this point in the development of the field of rebound economics, practical considerations dictate that economists seeking to measure or use such a production function limit themselves to a less general formulation. A practical approach is to confine the analysis by means of two key assumptions. First, one assumes the expression for the production function takes the following form:

$$Y = f(K, L, E, M, \ldots, \boldsymbol{\beta}, \tau_K, \tau_L, \tau_E, \tau_M, \ldots)$$

That is, the parameters $\boldsymbol{\beta}$ are assumed not to be affected by new technology. The second assumption is the technology parameters $\boldsymbol{\tau}$ are taken to be given 'exogenously' (and so do not depend on prices or anything else) and are specific to each input. They can be allowed to each move independently of one another over time. A more transparent expression of the above production function then looks like this:

$$Y = f(\tau_K K, \tau_L L, \tau_E E, \tau_M M, \ldots, \boldsymbol{\beta})$$

where the technology parameters simply multiply the inputs directly. This is the expression used throughout this chapter – although we forgo explicitly indicating $\boldsymbol{\beta}$ and treat these parameters as part of the description of f. While not entirely general, the formulation we use here gives us an easy means to translate concrete engineering efficiency concepts, expressed in the language of the engineer, into economic concepts expressed in the language of the economist.

This 'new' heat pump technology delivers more space heating/cooling using far less fuel than other systems of comparable capacity available on the market. How can we depict this relationship between the new energy efficiency and the old?

First recall that before this new technology came on the scene it took a certain amount of fuel input to the old system to produce the required level of comfort, a certain number of BTUs per year of natural gas, say. This new technology can accomplish the same thing with fewer BTUs of gas. It can, in terms of our production function, supply the same amount of comfort in the production of comfortable space Y using a smaller quantity of E. Said another way, this new smaller quantity of E is just as 'effective' in creating output Y as is the old, larger quantity. We can depict this with the following equation where τ is some multiplier greater than one:

$$E_{Effective} = E_{Old} = \tau E_{New}$$

This equation says if you wish to provide the same heating/cooling delivery to your building as you did with the old technology, it will require only half the fuel if $\tau = 2$. It also says the new technology is twice as 'effective' in delivering heating/cooling from every BTU of fuel it consumes. To reduce the notational burden, we simplify this equation to the following:

$$E_{Effective} = \tau E$$

This indicates that the reduced amount of actual physical energy input E needed with the new technology (BTUs for fuels, kWh for electricity) is just as effective in providing climate control as the greater amount of physical energy input used with the old technology (which was in fact τ times greater than the input needed now).

It is critical to realise that τ does not, or at least need not, depict thermodynamic conversion efficiency. Our heat pump is a good example. The heat pump delivers more space heating/cooling with less fuel than, say, a conventional furnace, but it does not do this by improving thermodynamic efficiency. It's just really good at 'pumping' heat from one reservoir to another (see Box 5.2). The more general point is that τ must

Box 5.2 Heat pumps

Here we elucidate the claim that a heat pump delivers more space heating/cooling with less fuel than a conventional furnace, but not by improving thermodynamic efficiency.

A heat pump is actually very inefficient thermodynamically (less thermodynamically efficient than a boiler) because what it does is take the heat content of the primary fuel and convert it to work. Converting heat to work is highly inefficient thermodynamically – much less efficient than just burning the fuel to produce heat directly. But what the heat pump does is use the work to 'pump' heat from one reservoir to another (e.g. from the outside of a building to the inside – or the reverse, if it is operating in cooling mode).

A heat pump can overcome the thermal bottleneck imposed by the second law of thermodynamics for the purpose of space heating. The ratio of energy transferred (the energy services) to the energy used in the process (some people call this the 'coefficient of performance (CP)') can exceed 100 per cent. In fact a typical CP for a commercial heat pump is between 3 and 4 units transferred per unit of energy supplied. *This means a heat pump can deliver more heat to a house than burning the primary fuel at 100 per cent efficiency inside the house.*

For typical air conditioners, the range in CP is 1.6 to 3.1. Refrigerators use similar engineering concepts. So the introduction of, say, a heat pump in place of a furnace could easily change the ratio of energy service provided (space heat, S) to fuel consumed (E) by 200 per cent. In other words, S/E doubles with a *reduction* in thermodynamic conversion efficiency. Said another way, the heat pump delivers the same 'effective' energy as the furnace with half the physical fuel input if $\tau = 2$, even though it is less thermodynamically efficient. So τ certainly does not represent thermodynamic conversion efficiency (or even a gain in thermodynamic conversion efficiency, necessarily).

comprehend all the engineering process changes that convert physical fuel into 'effective' energy (see later section, 'Technology parameters: examples and assessment'). In this sense, it is more an economist's version of technology than an engineer's. In the language of the economist, τ is called a 'factor-augmenting' technology gain. Nonetheless, we submit that it embodies concepts that are very familiar to the engineer.

We can view the relationship between 'effective' energy and capital in a way similar to Figure 5.1 (see Figure 5.2).

In Figure 5.2, we plot 'effective' energy against capital instead of energy against capital as was done in Figure 5.1. If the actual energy price stays fixed, an increase in τ, because it provides the same amount of 'effective' energy ($E_{Effective}$) and delivers the same output (Y) as can be delivered with less actual physical energy input (E) results in an 'effective' reduction in the energy price. Since the actual energy input (E) is now less than before the technology gain, it now requires less expenditure on actual energy to deliver the same output (Y). In other words, the price per unit of 'effective' energy delivered is less than before. We can think, therefore, of the effect being one of a reduction in the 'effective energy price'. As with Figure 5.1, a reduction in the effective energy price corresponds to a shift in the straight line to a new equilibrium point (black dot) and away from its previous equilibrium point (grey dot).

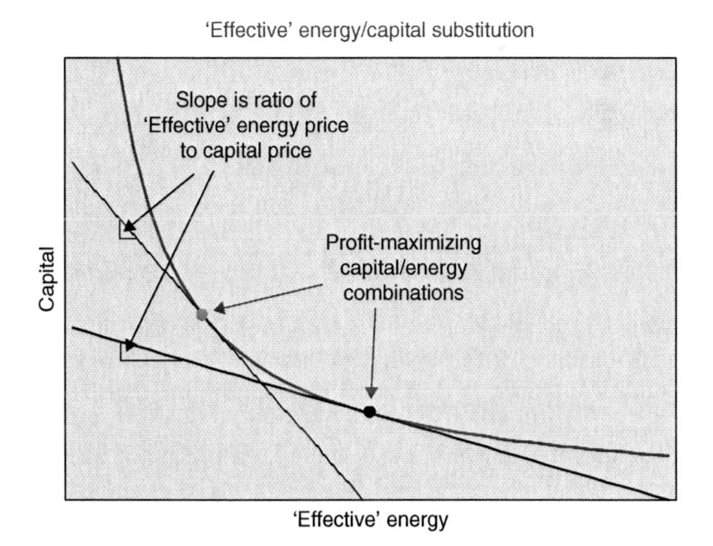

Figure 5.2 'Effective' capital-energy substitution

The result is an increase in demand for 'effective' energy. But since actual energy and 'effective' energy are locked together by the above equation, the result is also a tendency to increase demand for actual energy (E). Hence, while less actual energy is required for each unit of effective energy ($E/E_{Effective} = 1/\tau$), more effective energy will be used to produce a unit of output. It is an empirical question as to which of these two effects dominates and hence whether more or less actual energy will be used to produce a unit of output. Moreover, it will later be shown that this is only part of the effect of an increase in τ. A secondary effect is that the curved line will shift outward, indicating that the level of Y will be increased by an increase in τ. Since an increase in Y calls for an increase in inputs, this further increases demand for actual energy, E. Another way to say this is that an increase in τ not only decreases the effective energy price, but also increases the space of production possibilities. Both cause an increase in actual energy demand. (This graphical analysis was developed by Sorrell and Dimitropoulos (2007) and is explained in richer detail there.)

With this idea in hand, we can now redefine the production function to include engineering efficiency gains for all inputs:

$$Y = f(\tau_K K, \tau_L L, \tau_E E, \tau_M M, \ldots)$$

where τ_i is the factor-augmenting technology gain for the ith input.

Rebound: engineering intuition and production function analysis

Although we now have this more general production function at our disposal, let us return to the case where the engineering efficiency gain is limited to energy:

$$Y = f(K, L, \tau_E E, M, \ldots)$$

The existence of the new technology, the heat pump, changes your decision space as an engineer. You now are faced with again deciding the proper combination of insulation with an appropriately sized heat pump system. The situation is not as straightforward as simply replacing the system you selected before with a new heat pump of comparable capacity. You need to re-do the analysis to see what the new least-cost insulation/heat pump combination will be. You may decide, for example, that the best thing to do is to increase the capacity of your plant and remove some insulation from the program. If this is the case, the fuel use you contemplate would not be simply half of what it was with the old technology you proposed (if $\tau = 2$), but something greater than this. This is what economists call 'rebound' – it is less than a one-for-one reduction in fuel use compared to the fuel efficiency gain. A simplistic analysis would say you would decide to reduce your fuel use by half; the cost-minimising solution may say you would reduce it by less than that.

In fact, the price considerations may be such that you would decide to *increase* the capacity enough (and reduce insulation enough) that the fuel use you contemplate actually is *greater* than it would have been with the old technology. This is a condition economists call 'backfire'.

Now let us translate this into economic terms. Economists begin with a depiction of the optimisation problem you face. It looks like this:

$$\max p_Y Y - (p_K K + p_L L + p_E E + p_M M) \tag{1}$$

If p_Y is the price of the product your enterprise produces (we could think of lease prices for the space you are heating/cooling), Y is the amount of product you produce (square metres of climate-controlled space provided), and p_K, p_L, p_E, p_M are the prices of the inputs required to produce it, this equation simply says you will choose the amounts of K, L, E, M you apply to this enterprise in such a way as to maximise the difference between your revenue and cost – you will maximise your profit, in other words. If you think of your output price p_Y as fixed and given, and the quantity of output Y you will produce as fixed and given, this problem is the same as minimising cost.

Now, the way the economist will look at this problem is mathematically. We have presumed Y is characterised by the function f (later we'll come to what this function might look like). It turns out that this optimisation problem, when solved analytically, delivers several conditions – one for each input, in fact – but we want to look at the condition that relates to the quantity E, since we're interested in how this might change within the framework of the optimisation problem when E is replaced by its 'effective' equivalent $\tau_E E$.

The condition on E (called the 'first order condition') looks like the following:

$$\frac{p_E}{p_Y} = \frac{\partial Y}{\partial E} = \frac{\partial f(K, L, \tau_E E, M)}{\partial E} \tag{2}$$

This condition says profit will be maximised when the actual energy price p_E (as a fraction of the output price p_Y) is equal to the first derivative of the function f with respect to E. This derivative is referred to as the 'marginal productivity of energy', since it describes the incremental change in Y we get when we increase E a little bit. From this relationship, you can see how we can in principle solve for E in terms of the other elements of the equation. The final step in evaluating the rebound effect is to then, from this equation, derive how E will change in response to changes in τ_E. In a later section, we give an example.

Using production functions: policy analysis and prediction

Suppose we have our production function, we have had some econometrician measure its parameters, and now want to ask what to do with it. There are fundamentally two uses of this function for analysing rebound effects.

The first use of such a function is as a device for predicting the effect on energy consumption of some new or prospective technology. Policymakers may want to know this for purposes of allocating R&D resources to technologies that promise to reduce fuel consumption as a way of mitigating global warming, for instance. To do this, they will have to understand the way this new or prospective technology affects each of the inputs. In the following section, we give examples of such technologies and show how it is possible to assess the τ_i parameters in a practical way given an accommodating engineer or two.

The second use of such a function is for historical analysis and prediction. If one has measured the parameters of a production function including technology terms for, say, an entire economy over a certain historical period, one can evaluate how energy use would have evolved in the absence of that technology. This may give insights regarding the extent to which technology improvement has decreased or increased energy use. Such an analysis can also predict the future course of energy use assuming these technology trends continue (and the non-technology parameters of the production function remain fixed).

Technology parameters: examples and assessment

First let us consider a number of examples of new technology aimed at reducing energy use. Then we will outline a methodology for assessing the technology parameters associated with a particular technology at an application level.

Examples

Example 1: waste heat recovery. This example is one that, unlike our heat pump example, involves a straightforward increase in thermodynamic efficiency. Many industries have found that, with the increases in fuel prices seen over the past few decades, it is economical to put in place waste heat recovery systems. Some of the technology for doing this was previously unavailable and was developed in response to the increased fuel prices. Implementing these technologies usually involves the expenditure of capital, so both capital and energy are affected. In terms of our technology parameters, we would say τ_K has decreased (i.e. more capital is required to provide a unit of output) while τ_E has increased (i.e. less energy is required to provide a unit of output). In some cases the reduction in energy consumption may be substantial, but it is ultimately limited by the second law of thermodynamics.

Example 2: petrochemical production. The petrochemical industry is known for ongoing improvements in production processes. It is not uncommon for innovations to occur that reduce both fuel use and capital expenditures. Engineers do this by such means as changing reactor design, chemistry, and process flow design. For us, this would be characterised as increases in both τ_K and τ_E (i.e. reductions in both the amount of capital and the amount of energy required to produce a unit of output).

Example 3: plant reconfiguration. Sometimes it is possible to decrease fuel use simply by doing things smarter. Examples exist of plants where fuel consumption was reduced simply by reconfiguring assembly work stations and changing the flow of component parts to require less movement along conveyor belts. This is as close to a 'pure' energy efficiency gain as it probably gets; it is one we would characterise as an increase in τ_E alone.

Example 4: computers. If one considers the energy used by an old mainframe computer (climate controlled rooms with several operators and a maintenance crew, and so on) to produce a floating-point operation or a byte transfer and compares this with that used by your laptop, the gains in energy efficiency have clearly been enormous. Here we could think of the output Y as being a floating-point operation or byte transfer, and E as the corresponding power required to run the 'clock' and the hard drive, the fan to cool the board, and (if you think about Y as including making the calculations visible) the display. It is easy to see that the explosion of output from the computing industry has carried with it substantial improvements in capital efficiency, energy efficiency and labour efficiency. In technical terms, increases in τ_K, τ_L, and τ_E have been very large in the past few decades in this industry.

Assessing technology parameters

Let us imagine I am an energy policy analyst and you are the lead engineer on a team that has developed a highly innovative new production process for the pulp and paper industry. Your intellectual property claims on this new process are minimal because, while clever, it is based on knowledge in the public domain so you anticipate that when competitors see what you have done they will soon follow with copycat processes. For my part, I wish to understand how this new process will change energy demand in your industry.

Policy analyst: 'So, you say you have been able to eliminate several key steps in the process and so will spend less on capital equipment. How much less capital will you spend for this new plant compared to what you would have spent on a plant using the old technology?'

Engineer: 'Actually, we'll spend more because we'll make this plant much larger than we would have made an old-style one.'

Policy analyst: 'Okay, let's say you size it the way you think best. So tell it to me percentage-wise: how much less capital per ton of output product will you spend?

Engineer: 'Let's see, I calculate about 35 per cent less.'

Policy analyst (writes down $\tau_K = 1/(1 - .35) = 1.54$): 'And will you save any labour, using the same percentage-of-output basis?'

Engineer: 'Well, we'll use the same number of operators as we would for the smaller plant. Since the plant will be 50 per cent larger, I guess that means we'll use two-thirds of the labour.'

Policy analyst (writes down $\tau_L = 1/(2/3) = 1.5$): 'And what about energy? How much less fuel will you use?'

Engineer: 'This is one place we expect to make major gains. We anticipate we'll be able to cut our fuel use in half relative to output.'

Policy analyst (writes down $\tau_E = 1/.5 = 2$): 'And materials? Any gains there?'

Engineer: 'With the process configured as it is we should have a little less pulp wastage. I'd say about 5 per cent less.'

Policy analyst (writes down $\tau_M = 1/.95 = 1.05$): 'Good. Thank you.'

The policy analyst, armed with this assessment, goes back and develops a prediction for five years hence of the degree to which this new technology will penetrate the pulp and paper industry and develops a technology vector for this sector that can be inserted in a production function.[1] He then enlists an econometrician to measure the production function for this sector. With this, he can gain insights about the evolution of energy demand in the sector given this new technology.

Quantitative example of a rebound calculation

At this point it is useful to show how a technology vector such as that developed in the preceding section can be used to evaluate the change in energy demand it produces. This involves appealing to a production function specification and using the expression shown in Equation (2).

A simple production function that honours all the normal requirements economists insist on is the Cobb-Douglas production function. It turns out that this particular function is not 'rebound-flexible', that is, it is not suitable for serious rebound analysis, but because it is simple it allows for an easy illustration of how economists can evaluate rebound effects. The production function (ignoring materials) looks like this:

$$Y = aK^{\alpha}L^{\beta}E^{1-\alpha-\beta} \tag{3}$$

With our technology terms, it looks like this:

$$Y = a(\tau_K K)^{\alpha}(\tau_L L)^{\beta}(\tau_E E)^{1-\alpha-\beta} \tag{4}$$

For a Cobb-Douglas production function, the exponents $\alpha, \beta, 1-\alpha-\beta$ are equal the value shares (prices times quantities) of K, L and E, respectively.[2] Let's say we know the value shares to be $\alpha = .3$, $\beta = .6$, $1-\alpha-\beta = .1$.

By solving Equation (2) for E and plugging Equation (4) into the result, with a little algebraic manipulation we can develop an expression for E that does not depend on itself:

$$E = \left[(1-\alpha-\beta)\frac{p_Y}{p_E}a(\tau_K K)^{\alpha}(\tau_L L)^{\beta}\tau_E^{1-\alpha-\beta}\right]^{\frac{1}{\alpha+\beta}} \tag{5}$$

Table 5.1 Effect of different technology combinations on energy use and output

τ_K	τ_L	τ_E	E	Y
1.00	1.00	1.00	1.00	1.00
1.00	1.00	1.20	1.02	1.02
1.20	1.00	1.00	1.06	1.06
1.00	1.20	1.00	1.13	1.13
1.20	1.20	1.20	1.22	1.22

By similar means we can develop an expression for Y:[3]

$$Y = [a(\tau_K K)^{\alpha}(\tau_L L)^{\beta}\tau_E^{1-\alpha-\beta}]^{\frac{1}{\alpha+\beta}}\left[(1-\alpha-\beta)\frac{p_Y}{p_E}\right]^{\frac{1-\alpha-\beta}{\alpha+\beta}} \tag{6}$$

With these expressions, we can calculate what happens to E and Y when any of the technology terms change. More correctly, we can calculate what happens to them if we make some assumptions about K, L and the ratio p_Y/p_E. To illustrate, let us take a simple case where we hold these fixed. (Fixing the ratio p_Y/p_E, or more precisely, fixing its inverse, is what is called keeping the energy price p_E fixed 'in real terms'.)

Having done all this, we can generate predictions of what will happen to E (and Y) given different values of the technology parameters, as shown in Table 5.1.

From Table 5.1 it is possible to discern that a rebound effect for energy consumption arises in part because an increase in output 'drags up' energy use. A second, less obvious, contributor is that any efficiency gain that reduces the amount of energy needed to produce a given output (i.e. any increase in τ_E) thereby reduces the effective price of energy. As a result, effective energy will substitute for other inputs in the production of output, carrying up actual energy use with it, as we saw in connection with Figure 5.2. You can perhaps now see why economists need general functions that accommodate the substitution possibilities given changes in input prices, even if these are only changes in 'effective' prices.

So rebound occurs for two reasons: one, technology gains increase the space of production possibilities and so stimulate energy consumption; and two, these gains also reduce energy's effective price, causing it to become more attractive to use relative to other inputs.

Note that technology improvements in inputs other than energy cause a rebound effect, too. This is a fairly general finding in rebound theory, and means improvements in the productivity of other inputs, even if they are the by-product of engineering efficiency gains targeted at energy, can contribute to rebound.

Note also that, with the production function used here, an improvement in the efficiency of *any* input results in an outright *increase* in energy consumption compared to what it would have been without the improvement. This indicates such technology improvements will 'backfire', in rebound terminology. However, this is just a quirk of having chosen a Cobb-Douglas production function to illustrate the methodology; this particular function will always 'backfire', even for 'pure' energy improvements alone (i.e. increases in τ_E). Cobb-Douglas is not 'rebound-flexible' in that it cannot represent the full range of possible rebound effects – which theoretically includes 'super-conservation' (a greater than one-for-one reduction in energy use with engineering efficiency gain). Hence it is unsuitable for rebound analysis. But it shows us the idea.

Using derivations like this, it is possible to come up with formal mathematical definitions of rebound.

Finding and evaluating production functions

The example in the preceding section proves that we need to take care in selecting a production function for rebound analysis. Such a function needs to be rebound-flexible, it needs to honour the conditions economists like to see, it needs to easily allow the introduction of engineering-based technology parameters, and it needs to be measurable from real economic data. Finding such functions is not an easy task, but it is not impossible. We describe three candidates in the Annex.

The task of estimating the parameters of such functions is also challenging. In Saunders (2007) a methodology is outlined, which differs in certain respects among different production functions. That article is aimed primarily at economists and econometricians, but the engineer should take comfort in the fact that such an analysis is possible.

Conclusion

It is not presently known the extent to which rebound phenomena occur in reality (see, in particular Chapters 2 and 12). But the task of trying to measure whether they do, and if so to what degree, must begin from a foundation that honours both engineering principles and economic principles. The intent of this chapter is to begin presenting the case that such an endeavour is both possible and practical.

Annex: production functions for use in rebound analysis

There are many candidates for the function f. Some can be excluded as being too inflexible to use for rebound analysis (Saunders, 2007). Here we discuss three 'rebound-flexible' functions that can be specified in a form allowing econometric measurement of their parameters, including the technology parameters developed according to the 'engineering efficiency' methodology of this chapter. Each function has advantages and disadvantages.

For reasons related to the ability to 'aggregate' results, economists find it more useful to measure not the production function but its mathematically equivalent 'dual' cost function. For the three functions described below this methodology is described in Saunders (2007). The reader can be assured that the results so developed directly translate into the production function approach we have used to describe engineering efficiency gains in the main body of the chapter.

Measuring such a function requires using historical data, meaning the analyst must accommodate trends in technology, not just one-time changes. For us, practicality dictates we make the assumption that technology change is uniform with time, so that for the ith input: $\tau_i = \overline{\lambda}_i e^{\lambda_i t}$. Econometricians will be able one day to relax this assumption and measure period-to-period changes in technology.

The functions we discuss are the Symmetric Generalised Barnett, the Gallant (Fourier), and the Translog. Each of these is examined in more technical detail in Saunders (2007). We take each of them in turn.

Symmetric Generalised Barnett

The Symmetric Generalised Barnett (SGB) function is due to Diewert and Wales (1987) and is based on an original form developed by Barnett (1982). In the form we use here, it is equivalent to the Generalised Leontief cost function, developed earlier by Diewert (1971). The key advantage of the SGB is that it is globally concave, meaning additional restrictions do not have to be applied to estimate it thereby greatly simplifying the measurement task. ('Concave' is the restriction referred to in the main text that the more you try to favour one input over others the harder and harder it should become to produce gains in output.) It is also relatively flexible in depicting rebound.

There are two disadvantages of this form. First, it is not completely 'rebound flexible' since it cannot depict 'super-conservation'. By this, we mean the theoretical, if somewhat counterintuitive possibility that a 1 per cent increase in fuel efficiency may result in a greater than 1 per cent reduction in energy use (Saunders, 2008). Second, the factor-augmenting technology specification we use cannot be entered into it exactly, so we must rely instead on a method that introduces an approximation.

Gallant (Fourier)

This function is an extremely elegant one due to Gallant (1981), based on a mathematical form called the Fourier series. The advantages of the Gallant (Fourier) function are that it is fully 'rebound flexible' and it is completely general, in principle, in its ability to statistically emulate the 'real' functional form presumed to underlie the form being measured. The disadvantages are that it requires additional restrictions to force concavity and, like the SGB, the factor-augmenting technology specification we use cannot be entered into it exactly thus forcing us to rely on a method that introduces an approximation. A further impediment to its use is that, even given this technology approximation, it is inherently difficult to enter technology into the function, thereby requiring more complex estimation procedures.

Translog

This function was introduced by Christensen et al. (1971) and is used widely within energy economics. The key advantage of the Translog is that it allows our technology specification to be entered in an exact, as opposed to approximate form, unlike the two functions above. By accommodating this, it also enables deep insight into the possible limitations of the factor-augmenting, constant-technology-gain model.

The disadvantages of this form are that it requires additional restrictions to force concavity and it is not fully 'rebound flexible'. In fact, if global concavity restrictions are imposed on it this form is completely rebound inflexible, depicting only backfire. Ryan and Wales (2000) have found a way to introduce concavity locally in a way that allows this function to be *relatively* rebound flexible (it still cannot depict super-conservation), but the methodology requires concavity to be checked *ex post*, and it is not clear what remedy to apply if it is not.

Acknowledgements

I would like to acknowledge Steve Sorrell for extensive and fruitful discussions on this topic and on the subject of rebound generally. This dialogue has been critical in shaping my thinking and the content of this chapter. Also, John Dimitropoulos provided important insights. Of course, all errors are my own.

Notes

1. Methods for doing this are found in Saunders (Saunders, 2005: Appendix B).
2. With technology, they are actually the value shares of 'effective' capital, labour, and energy, respectively, that is of $\tau_K K$, $\tau_L L$, $\tau_E E$.
3. Taoyuan Wei (2006) developed this derivation.

References

Barnett, W. A. (1982) *The Flexible Laurent Demand System*, Proceedings of the 1982 American Economic Association Meetings, Business and Economics Statistics Section: 82–9.

Christensen, L. R., D. W. Jorgenson and L. J. Lau (1971) 'Conjugate duality and the transcendental logarithmic production function', *Econometrica*, 39: 255–6.

Diewert, W. E. (1971) 'An application of Shephard Duality Theorem: a generalized Leontief production function', *Journal of Political Economy*, 79: 481–507.

Diewert, W. E. and T. J. Wales (1987) 'Flexible functional forms and global curvature conditions', *Econometrica*, 55: 43–68.

Gallant, A. R. (1981) 'On the bias in flexible functional forms and an essentially unbiased form: the Fourier flexible form', *Journal of Econometrics*, 15: 211–45.

Ryan, D. L. and T. J. Wales (2000) 'Imposing local concavity in the Translog and Generalized Leontief cost functions', *Economics Letters*, 67: 253–60.

Saunders, H. D. (2005) 'A calculator for energy consumption changes arising from new technologies', *Topics in Economic Analysis and Policy*, 5(1): Article 15. See: http://www.bepress.com/bejap/topics/vol5/iss1/art15

Saunders, H. D. (2007) 'Econometric specification of three rebound-flexible production functions', manuscript available from the author (hsaunders@decisionprocessesinc.com).

Saunders, H. D. (2008) 'Fuel conserving (and using) production functions', *Energy Economics*, forthcoming.

Sorrell, S. and J. Dimitropoulos (2007) 'UKERC review of evidence for the rebound effect: Technical Report 5: Energy, productivity and economic growth studies'. See: http://www.ukerc.ac.uk/Downloads/PDF/07/0710ReboundEffect/0710TechReport5.pdf

Wei, T. (2006) 'Impact of energy efficiency gains on output and energy use with Cobb-Douglas production function', *Energy Policy*, 35: 2023–40.

6

Energy Efficiency and Economic Growth: the 'Rebound Effect' as a Driver

Robert U. Ayres and Benjamin Warr

In this chapter we argue for two linked theses. The first thesis is that increasing energy efficiency has been a major – perhaps the major – driver of economic growth since the industrial revolution. In fact, 'technological progress' as normally construed by economists appears to be primarily due to increasing energy-conversion efficiency, notwithstanding contributions from information technology in recent decades. The second, related thesis is that, while reduced carbon dioxide emissions are essential for long-term global sustainability, the usual policy recommendation of most economists and many 'greens', namely to increase the cost of energy by introducing a carbon tax, would be ill-advised. Such a policy might have the beneficial effect of reducing government budgetary deficits. But it would have an adverse impact on economic growth, at least in the industrial world.

In respect to the first of our two theses, we note that the standard theory assumes that economic growth is automatic, inevitable and cost-free, and on an optimal path. It follows that growth will continue in the future at essentially the same rate ('the trend') as it has in the past. Some economists have concluded, on the basis of this preconception, that government interference – e.g. to ameliorate climate warming – is more likely to inhibit growth than to promote it (Nordhaus, 1991, 1992). Another misleading implication of the 'trend' assumption is that our children and children's children will inevitably be much richer than we are. It would follow that future generations can better afford than this generation to repair the environmental damages that past generations have caused (Schelling, 1997). We disagree with both the underlying assumption of automatic growth and with the 'do nothing' policy implications.

But the second thesis also has serious policy implications. To the extent that economic growth has been driven in the past by increasing energy-conversion efficiency, it is important to find ways to keep that trend going as far as possible. But efficiency has limits, so it is also important to promote alternative technological drivers with broad impacts on other sectors of the economy. Information technology appears to be the only plausible candidate at this time.

Economic background

Almost nobody (with the possible exception of a trained neoclassical economist) would doubt that energy is as essential to the functioning of the global economic system as gasoline is to a car or electricity to a light bulb. In a 1968 article, V. E. McKelvey, Director of the US Geological Survey, began with a quotation:

> Our civilization ... is founded on coal, more completely than one realizes until one stops to think about it. The machines that keep us alive, and the machines that make the machines are all, directly or indirectly, dependent upon coal ... Practically everything we do, from eating an ice to crossing the Atlantic, and from baking a loaf to writing a novel, involves the use of coal, directly or indirectly. For all the arts of peace coal is needed. If war breaks out it is needed all the more.
>
> (Orwell, 1937, cited by McKelvey, 1968)

McKelvey went on to say:

> To make Orwell's statement entirely accurate – and ruin its force – we should speak of mineral *fuels*, instead of coal, and of other minerals also, for it is true that minerals and mineral fuels are the resources that make the industrial society possible.
>
> (ibid.)

The evidence is visible and pervasive. But to reiterate and emphasise the obvious, it is the sun's energy that drives the most fundamental process in nature, photosynthesis, whereby carbon dioxide and water are converted into carbohydrates and cellulose – or, more familiarly, biomass. The biomass produced in the distant past was converted by natural processes into the fossil fuels we utilise – wastefully – today. Meanwhile, it is the fossil residue of past photosynthesis that drives the world's factories and transport systems, heats its homes and generates much of its electricity.

Nevertheless, macroeconomists traditionally underestimate the importance of energy. Standard textbooks include such curious illustrations as the theoretical bakery that employs capital and labour (both rented from some external agency) to produce bread, without either flour or fuel.[1] The logical consequences of this fundamental impossibility have been put to a very misleading use, in the mathematical specification of something called a 'production function'. The neoclassical production function was introduced into economics long ago to quantify a formerly qualitative understanding. In brief, it is supposed to explain an output, namely economic activity (and growth), in terms of input variables: namely the two so-called 'factors of production', capital and labour. How did it happen that energy got left out of the standard theory despite its overwhelming importance in the real world? Part of the answer is that the importance of energy was not nearly so obvious to early economists like Adam Smith and Joseph Ricardo. Both the products of photosynthesis and, of course, coal, were associated with a third factor of production: land. However, one would have to recapitulate the intellectual history of economics to fully elucidate this question.[2]

To be sure, economists have not been able to completely ignore energy, especially since the events of the 1970s. Again, this is not the place to review the history of attempts to incorporate energy explicitly as a variable into the standard production function approach.[3] Suffice it to say that the first attempts were not successful in the sense of explaining past economic growth. In retrospect, it appears that this failure was primarily because they began with two doubtful assumptions. The first is known in the trade as 'constant returns'. In simple language this means that a large economy is not *ipso facto* more productive (per capita) than a small one. The large size of the US in relation to Sweden or the larger size of China in relation to the US does not give the bigger country an automatic advantage in terms of productivity per capita. This assumption is generally accepted, despite the obvious fact that economies of scale do play a significant role at the firm level and at the community level.

The second, more questionable assumption incorporated in many economic models is that the output elasticities[4] (or marginal productivities) of the input variables must be reflected by the allocation of payment shares in the national accounts. It has been convenient to divide all payments into two categories, namely wages and salaries (returns to labour) and dividends, royalties, interest and rents (returns to capital). There is a theorem to the effect that, if firms are competitive price takers and cost minimisers, and if all firms produce a single composite product, then payments to a factor will be allocated according to the marginal productivity of that factor. But energy does not appear explicitly on the payments side of the national accounts, although payments to 'energy' can be roughly equated to revenues of certain industries, such as coal mining, petroleum and gas drilling If this is done, it turns out that the energy 'share' of payments in industrial countries is negligible, not more than a few per cent – around 4 per cent of the US GDP, for instance (before the recent increases in energy prices). Most economists in the past have assumed, despite intuition to the contrary, that (due to the income allocation theorem mentioned previously) energy cannot be very 'productive' as compared with capital or labour.

The standard escape from this dilemma has been to assume that energy is not really a primary variable (i.e. a factor of production) but, rather an *intermediate* product of the economy. The argument is that capital and (human or animal) labour – combined as in the construction of a watermill, or digging a coal mine – produce the materials and energy 'intermediate goods' that are subsequently converted into products and final services. This explanation seems plausible at first sight, and has rarely been challenged. However, it ignores the fact that every process and activity in the economy requires energy to function. More precisely, they require 'available energy' or 'exergy', which is explained further in the next section.

Labour and capital cannot account for the fundamental role of solar exergy as a direct input to agriculture, or as the source of natural forces like wind and falling water. The food or feed consumed by the human or animal workers is essentially pure exergy (calories). Moreover, exergy is *embodied* in most finished materials, like steel, aluminium or plastics, or in silicon chips, and thus in material goods. Finally, exergy is needed to transform raw materials into finished goods, to move them to consumers and finally to dispose of wastes. In fact, neither labour nor capital could exist without exergy.

Digression: on exergy, power and useful work

Before introducing an alternative perspective on the relationship between energy, economy activity and economic growth it is essential to clarify the distinctions between energy, exergy, power and useful work.

Fuel inputs are conventionally measured in terms of their heat content, or the amount of heat released when the fuel is burnt under certain standard conditions. But this does not take into account the 'availability' of energy, or the ability to perform *useful work*. 'Work' may be defined as energy that is transmitted from one system to another in such a way that a difference of temperature is not directly involved (Halliday and Resnick, 1966). Examples of work would be lifting a weight against the force of gravity or overcoming rolling or sliding friction. The more common term, *power*, is simply the rate at which useful work is produced by a *prime mover*, such as a steam-electric generator or an internal combustion engine.

A common measure of the ability to perform useful work is *exergy*, defined as the maximum amount of work obtainable from a system as it comes to equilibrium with a reference environment (and thus requires both the system and the reference environment to be specified) (Ahern, 1980). Exergy is only non-zero when the system under consideration is distinguishable from its environment through differences in either relative motion, gravitational potential, electromagnetic potential, pressure, temperature or chemical composition (Ayres, 1998). Unlike energy, exergy is 'consumed' in conversion processes, and is mostly lost in the form of low temperature heat. The essence of the second law of thermodynamics is that while exergy can always be turned into heat, heat generally cannot be turned into exergy.

Measuring energy inputs on the basis of their thermal content effectively implies that their ability to perform useful work is equivalent to their ability to generate low temperature heat. But this 'lowest common denominator' measure overlooks important qualitative differences between energy carriers. A heat unit of oil or natural gas will be ranked higher on an exergy basis than a heat unit of coal, since the former can do more useful work.

For present purposes, useful work can be divided into three major categories, which will (hopefully) also help to clarify the meaning. The first category consists of muscle work, performed by humans or animals. The most common example, historically, would be the use of horses for transport or farming. The second major category is mechanical work, which is mostly done by heat engines such as steam turbines, gasoline engines, diesel engines or gas turbines. Electric power is the output of a combination of a prime mover and an electric generator. Electric power can, in turn, be reconverted into mechanical work at some distant point by means of a motor, or it can be reconverted into heat (as in a toaster) or into chemical work (as in electrolysis). Hence electric power *per se* can be thought of as a sort of useful work, whereas the work done by a motor or electric heater is secondary. Finally the third major category consists of heat, whether at high temperature or low temperature, *delivered to a point of final use*.

We note that in common language usage, energy is *'converted'* into work. But the units we use to measure useful work (e.g. joules or kilowatt-hours) are also units of energy and exergy. This is a source of much confusion, and probably accounts for the fact that most economists have failed to distinguish clearly between energy, which is an input, and useful work, which is an output. However, the ratio of the output (useful work) to the input (energy or exergy) is a pure number, between zero and one. This ratio is known as the *efficiency* of conversion. The ratio of useful work to the heat content of fuel inputs is known as the 'first law' efficiency of the conversion process. But a more relevant measure is the second law efficiency, which compares the actual exergy used with the theoretical minimum required. This second law measure of energy efficiency is typically smaller than the first law efficiency, suggesting a great potential for improvement. As an example, the second law efficiency of conversion of heat energy from fuel to electric power delivered to consumers in industrial countries is roughly 33 per cent.[5]

Technological progress and efficiency

It is conventional wisdom that growth depends on 'technological progress'. But the term technological progress may be applied to Gillette's latest razor blade design or a detergent that washes shirts 'whiter' and nobody can claim that this sort of progress contributes significantly to overall economic growth. It merely replaces an earlier product and encourages waste. In fact, the vast majority of inventions and innovations are essentially small improvements in an established product, or the process by which it is made. Most economic growth is attributable to another sort of innovation, which has been called a 'general-purpose technology'. General-purpose technologies (GPTs) are defined by Lipsey et al. (2005) as technologies that: have a wide scope for improvement and elaboration; are applicable across a broad range of uses; have potential for use in a wide variety of products and processes; and have strong complementarities with existing or potential new technologies. GPTs reach across product boundaries and enable the creation of new products or services, or at least improve the performance and cut the cost of established products.

General-purpose technologies can be tools, such as the lathe, or new materials with new properties, or more efficient ways of producing, transmitting and utilising exergy to do useful work. There have been a few – and only a few – important general purpose technologies in the history of human civilisation. The wheel, the sail, the rope, the pot, the plough, the grinding wheel, the knife and the stirrup are among the oldest examples, along with swords, arrows and guns. The water wheel, the horse-drawn cart, the windmill and the oil lamp are more recent, but still very old.

We have already noted that economic activity is driven by available energy, or exergy, in various forms. More precisely, we argue now that it is *energy converted to useful work* that drives production, along with capital and labour as traditionally defined.[6] The intuitive reason why raw exergy inputs do not explain growth is that their inefficient conversion leads to a large fraction of waste heat (and other wastes, like ash) that do not contribute to the economy but actually create health problems and costs of disposal. We also note that, before the industrial revolution, most work was muscle work, with a tiny contribution from windmills, sails and watermills. Economic growth since then has been a long process of substitution of machines, driven mostly by the combustion of fossil solar energy,[7] for human and animal muscles. Fossil fuels also drive almost all of the metallurgical and chemical processes that create modern materials from steel to plastics.

The industrial revolution may have begun with cotton spinning and weaving, but it was powered by the steam engine – a paradigmatic example of a GPT. This machine was invented by Savery and Newcomen around the turn of the eighteenth century, originally to pump water from flooded coal mines, where it replaced expensive oat-consuming horses and treadmills. But it was improved dramatically by James Watt and soon applied to other purposes, including mechanised looms, boring machines, bellows to pump air into blast furnaces, newspaper presses and, of course, the railway and the steamship which revolutionised transport. Among other things, the steam engine made factories possible in places where there was no falling water. In effect, steam power did 'useful work' in many applications that had formerly required water power, animals or sails.

A century after Watt's major innovations came a new source of mechanical work: the internal combustion engine. Like Watt, there were precursors, notably Lenoir. But the milestone was Nikolaus Otto's commercial gas engine – using coal gas – designed to provide power for small factories or shops, in place of stationary steam engines. His assistant, Gottlieb Daimler saw that if the engine could be made smaller and burned a liquid fuel it would have automotive applications. Daimler invented the carburettor, concentrated on increasing speed and reducing weight, and joined carriage-maker Karl Benz. At the same time Rudolf Diesel began developing the more fuel-efficient compression-ignition engine, originally for ships, then railways and now automobiles.

Meanwhile, in the 1870s and 1880s Edison and Tesla (working for Westinghouse) introduced the core inventions of the electrical industry. Edison was originally concerned with lighting. He did not invent the dynamo (DC generator), but he doubled the efficiency of earlier versions. He did 'invent' the carbon-filament incandescent light (simultaneously with Swan in the UK). But his real innovation was the complete system for providing light (and heat) to a hotel or theatre or ship. This system found immediate applications in a host of areas having nothing to do with lighting – including trams and trains, elevators, electric arc furnaces and electrolytic cells. Soon after, Tesla and Westinghouse vastly improved the utility of electric power by introducing 3-phase AC – permitting efficient long-distance transmission – and AC induction motors.[8]

The series of pioneering inventions that followed Otto, Diesel, Edison and Tesla, along with the internal combustion engine, the automobile and the airplane, have together contributed much to the economic growth of the twentieth century. Of course, plastics, the transistor and the computer triggered a new wave of inventions and innovations that certainly contributed to economic growth in the second half of the twentieth century, and hopefully into the next. However, there is strong quantitative evidence to suggest that those two general-purpose technologies, heat engines and electric power, taken together, have been the 'growth engine' of the modern world for over a century (Ayres and Warr, 2005).

The efficiency-growth connection

What is the link between useful work output and economic growth? It is conceptually very simple, though much less simple to demonstrate quantitatively. In brief, as technological progress makes the conversion process from 'raw' exergy (e.g. fuels) to useful work more efficient, the cost of work tends to decline. As costs fall, in a competitive market, prices fall also. Declining prices of work generate increased demand for work, either for direct services (like heat) or for other goods and services that require inputs of useful work, as motive power, electric power or heat. This tendency is quantifiable as the price elasticity of demand. We have already noted that work and/or embodied exergy are required at every stage of the economic system from extraction to finished goods and final services. Increased demand, in turn, requires more investment and bigger production units with more economies of scale and also encourages learning-by-doing (increasing experience). This reduces production costs and product prices, thereby further increasing demand.

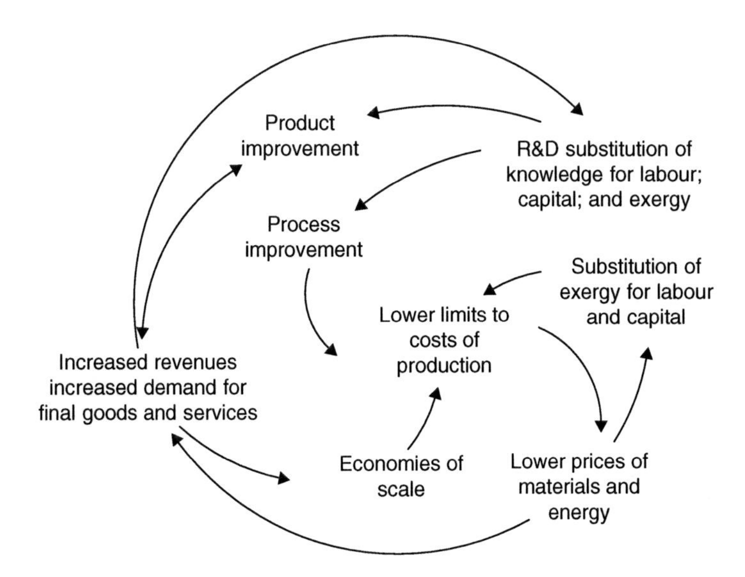

Figure 6.1 The extended Salter cycle of growth

Economists have generally simplified this relationship by reversing the implied causality, i.e. by assuming that investment drives growth. This is another way of expressing the idea that supply creates demand. In reality, supply and demand drive each other. In short, the growth 'engine' is a positive feedback cycle (Figure 6.1): mechanisms such as increasing conversion efficiency drive growth via declining prices and growth, itself, drives more growth.[9]

This cycle applies to all products, of course, but we think that the aggregate 'useful work' captures most of the important relationships: as the price of useful work declines the demand for it grows apace. Since it is, indeed, a cycle, it is equally correct to say that increased demand for useful work 'drives' economic growth, which – in turn – implies that increasing costs of useful work, or decreased supplies, could put the 'growth engine' into reverse, resulting in more unemployment, lower wages and so on.

The above verbal description of the past – and present – 'engine' of economic growth is essentially equivalent to the economy-wide 'rebound effect' as described by Jevons (1865), Brookes (1990) and Saunders (1992). However, the process can be demonstrated more formally through the mathematical model of economic growth developed by Ayres and Warr (2005). This departs from conventional models of economic growth in

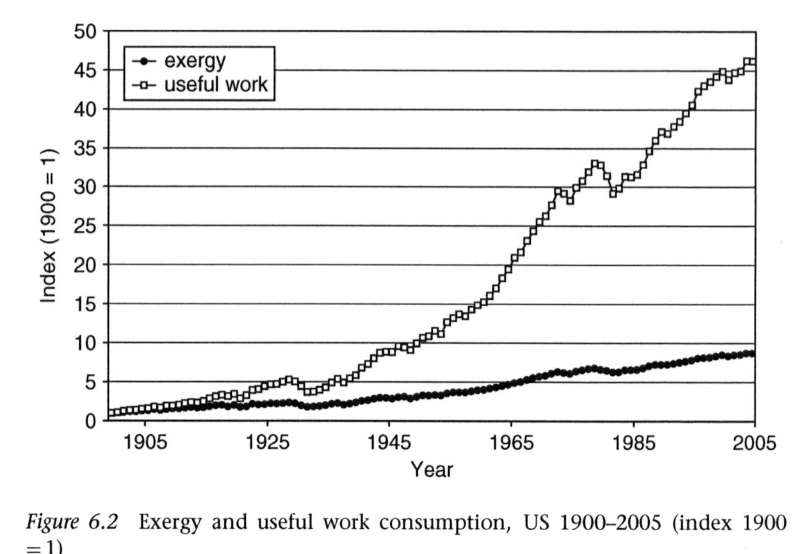

Figure 6.2 Exergy and useful work consumption, US 1900–2005 (index 1900 = 1)

two ways. First, instead of using energy (E) as an input to an aggregate production function, it uses *useful work* (U). This, in turn, is estimated from the product of exergy inputs to the US economy and a time series of second-law conversion efficiency, as derived by Ayres et al. (2003). As Figure 6.2 shows for the US, demand for useful work over the past century has increased at a much faster pace than the demand for exergy.

Second, it uses an unconventional form for this production function – namely a linear-exponential, or LINEX function – in which the marginal productivities of each input are assumed to be neither constant nor proportional to the share of that input in the value of output. The LINEX production function is obtained by choosing the simplest non-trivial solutions of the growth equation[10] and selecting plausible mathematical expressions for the output elasticities (α, β, γ) for each factor of production (K, L, U), based on asymptotic boundary conditions (Kuemmel et al., 1985). To satisfy the constant returns (Euler) condition these elasticities must be homogeneous zeroth order functions of the independent variables.

The first of Kuemmel's proposed choices can be thought of as a form of the law of diminishing returns (to capital). It is a boundary condition that conveys the notion that even in a highly capital intensive future economy, some labour L and work U are still required., viz

$$\alpha = a\frac{L+U}{K} \tag{1}$$

The second equation, for the marginal productivity of labour, reflects the continuing substitution of labour by capital and useful work as capital intensity (automation) increases:

$$\beta = a\left(b\frac{L}{U} - \frac{L}{K}\right) \tag{2}$$

Finally, the constant returns to scale condition defines the marginal productivity of useful work,

$$\gamma(K, L, U) = 1 - \alpha - \beta \tag{3}$$

Partial integration of the growth equation yields the LINEX function.

$$Y = U \exp\left[a\left(2 - \left(\frac{L-U}{K}\right)\right)\right] + ab\left(\frac{L}{U} - 1\right) \tag{4}$$

When this model is used taking capital, labour and useful work as inputs, it turns out that US and Japanese economic growth can be 'explained' quite accurately throughout the twentieth century *without* the need for an exogenous multiplier representing 'total factor productivity' (TFP) improvements, or technical change[11] (Figure 6.3). In conventional models of economic growth, this multiplier 'explains' a large proportion of the economic growth observed throughout the last century, with the remainder accounted for by increases in capital, labour and natural resource inputs. In effect, improvements in the thermodynamic conversion efficiency act as a suitable proxy for technical change, since these increase the amount of useful work available from each unit of energy input (Figure 6.4).

Policy implications of the new perspective on growth

The policy implications of this new perspective on growth are quite simple and stark. One major implication is that current policy prescriptions by most environmental economists and 'greens' are risky. The need to reduce consumption of fossil fuels and the emission of so-called greenhouse gases (GHGs) cannot be denied. But the usual prescription is to

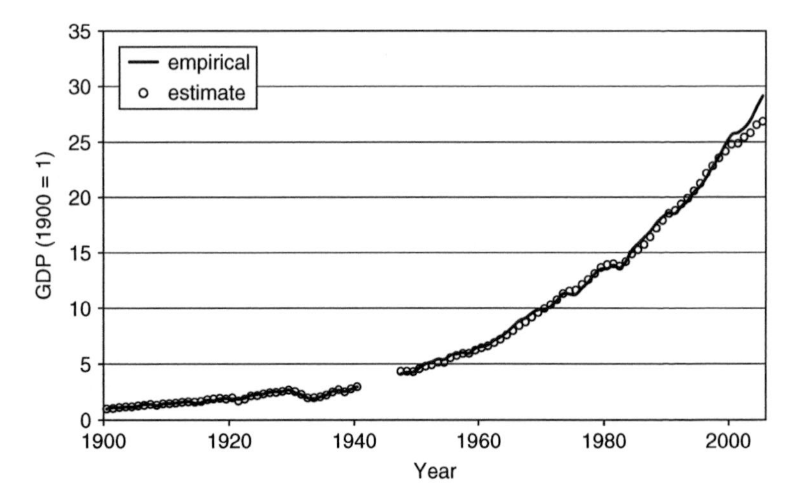

Figure 6.3 Empirical and estimated GDP for US 1900–2005

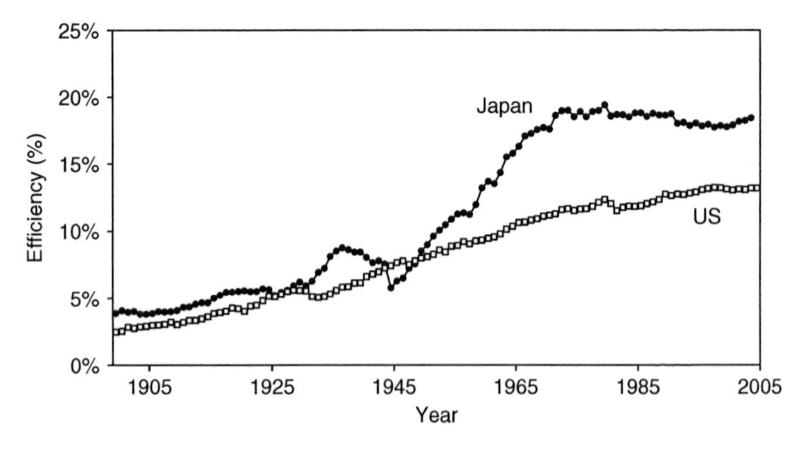

Figure 6.4 Aggregate efficiency of conversion of exergy to useful work for the US and Japan 1900–2005

'get the prices right' – e.g. to raise them – perhaps via a carbon tax, and thus to cut demand. If the standard textbooks are correct, higher energy (exergy) prices would cut energy consumption as required, but would have little effect on the economy as a whole. This is because exergy in the form of fossil fuels and hydropower accounts for such a small share – around 4 per cent – of the gross national expenditures. But from the perspective we have outlined above, *this prescription would put the mechanism of the economic 'growth engine' into reverse*. Higher prices would cut demand, not just for fuels but for downstream products and services at every stage. It follows that, to address the real problem of climate change, and to reduce the consumption of fossil fuels while at the same time encouraging continued economic growth, it is essential to focus simultaneously on three things: (1) to increase conversion efficiency, so as to produce more useful work with less energy (exergy) inputs, (2) to reduce fossil fuel exergy inputs in favour of zero-carbon alternatives and (3) to do so at lower cost. The real challenge lies in the third objective, as improving efficiency intrinsically leads to cost reductions as inputs are not wasted. Whether renewable exergy can be supplied in the quantities required at lower cost is debatable.

Evidently petroleum and gas are not going to become cheaper in the future, especially because demand from rapidly industrializing countries such as China and India is now overtaking traditional supply. Nor is nuclear power the panacea, owing to high costs, safety concerns and difficulties with waste disposal – although the lack of investment in nuclear power in recent decades has more to do with rising cost than public objections.[12] Moreover, electricity is not a direct substitute for liquid fuels, especially for mobile power.

But new and sophisticated technologies are unlikely to achieve this double objective either, at least in the near term. Hydrogen fuel cells are more efficient than gasoline engines, but the hydrogen itself must be ultra-pure and at present is obtained from hydrocarbons, yielding waste carbon dioxide and no obvious advantage in terms of carbon emissions. Carbon sequestration is technically feasible, at least for central generating facilities, but at significant added cost. Combined cycle generating plants are an attractive possibility, but only where natural gas is readily available. If the fuel must be obtained by coal gasification, the efficiency advantage is negligible and both costs and emissions are higher.

Solar PV (and to a lesser extent wind power) are not yet cost competitive, except in special and somewhat rare circumstances. The natural rate of market penetration is likely to be slow at best. These technologies are unlikely to become cost-competitive with central electric power generators unless *either* (1) the scale of production can be increased by orders of magnitude, *or* (2) the (mostly hidden) subsidies presently received by electric utilities are eliminated. The first option would require extensive subsidies or regulatory interventions to favour the newer technologies that would, themselves, introduce costs into the economy. The second option we address below.

What remains? The dominant policy option is to target increased conversion efficiency at lower cost. There may be significant possibilities to achieve 'energy conservation' at negative cost – Lovins has introduced the evocative term 'negawatts' – as applied primarily to end-uses.[13] The problem with that option is that the opportunities for achieving negative cost are limited. Those options should be exploited, without doubt. We think they are more widespread than the sceptics will admit. A good example of the potential is known as decentralised combined heat and power (DCHP), which offers significant savings in cost as well as reductions in GHG emissions (Casten, 1998; Casten and Downes, 2004; Ayres et al., 2007). One problem in this case is that decentralised power generation is strongly opposed by the incumbent electric power producers.[14]

The other alternative – and possibly the only one available in the longer term – is to promote more rapid innovation and adopt higher cost options more quickly, in order to bring down costs through economies of mass production (scale and learning). This approach is currently being adopted with respect to wind turbines and rooftop solar photovoltaic systems.

Conclusions

This chapter makes two points, which are worth repeating and emphasising. The first is that growth is driven by innovation but not all innovation is important. Most growth in the past two centuries has been driven by a series of general purpose inventions whose overall effect has been to increase the efficiency with which crude sources of exergy (including fossil fuels, but also food and agricultural products) are converted into useful work. But the historical trend in terms of efficiency increase has slowed down.

The second part of this chapter deals with the future. If we are correct in asserting that growth in the past century (or two) has resulted from the feedback cycle sketched previously, then it also follows that growth in the future is anything but certain. A minimal condition for continued economic growth would seem to be continued increases in the production of useful work, at lower cost and prices, by the economy. If new and cheaper resources are no longer to be found, the only available tool for achieving this result is to increase the efficiency of converting exergy to useful work. The question is whether this is still possible – given tougher environmental regulations and a forthcoming peak in global oil production, perhaps signifying the end of the 'age of oil' (Strahan, 2006).

The very first step, however, must be to recognise the importance of efficiency.

Notes

1. This typical example can be found in a popular recent textbook entitled *Macroeconomics* by N. Gregory Mankiw, a Harvard Economics Professor who is currently the Chairman of the Council of Economic Advisers to the President (Mankiw, 1997). The purpose of the example is to explain why the national accounts of an economy in equilibrium will pay capital and labour, respectively, in accordance with their respective productivities. Energy is not mentioned in the example, the theorem or the national accounts.
2. This is not the place for such an inquiry. However, for those interested, a good starting point is the 1989 article 'More heat than light' by Philip Mirowski (1989).
3. The 'KLEM' approach to modelling output growth, developed in the 1970s includes measures of energy (*E*) and materials (*M*) inputs, alongside capital (*K*) and labour (*L*). However, this model still rests on the assumption that improvements in total factor productivity are 'exogenous' – in other words they occur independently of changes in *K*, *L*, *E* or *M*.
4. As an example the mathematical formula of the elasticity of output (Y) with respect to capital K is simply $\partial \ln Y / \partial \ln K = (\partial Y / \partial K)(K/Y)$.
5. The astute reader may notice that if the electricity delivered is subsequently converted to a secondary service, like electric light (only 10 per cent efficient) the overall efficiency with which the economy produces the service of illumination is only 3 per cent. In fact, one of the authors has estimated that the US economy is currently operating at an efficiency (in the technical sense) of the order of 3 per cent, or less (Ayres, 1989). Clearly there is still a lot of potential for improved energy efficiency.
6. What most human workers do nowadays involves sensory inputs, eye–hand coordination, judgement and decisions, with very little use of muscle strength as such.
7. Hydroelectric and nuclear power are obviously also important in some countries.
8. Telecommunications and modern developments in electronics also benefited from another aspect of Edison's work, his 'Edison effect' (1883). This led to Fleming's vacuum tube diode (a rectifier), de Forest's triode (a detector or amplifier) and then to repeaters, radio circuits and in short, to radio telephones, long distance telephony, radio broadcasting, TV and computers.
9. We note, immediately, that economic growth before the industrial revolution was very slow, because the primary source of useful work – by human and animal muscles – could not be increased in efficiency. In other words the productivity of muscle work remained virtually constant. It follows that growth in GDP in every country is directly correlated with the mechanisation of agriculture and the movement of rural agricultural workers – no longer needed to work the soil – to towns and cities, where they became available to work in factories using machines.

10. The growth equation is the total (logarithmic) time derivative of the production function, viz

$$\frac{dY}{Y} = \alpha\frac{\partial K}{K} + \beta\frac{\partial L}{L} + \gamma\frac{\partial U}{U} + \delta\frac{\partial A}{A}$$

where the last term reflects the possibility that not all growth can be explained by K, L and U alone, and that other forms of 'technological progress' exist that cannot be directly quantified.

11. The parameters a and b in the LINEX function can be modelled as time dependent logistic functions. However both parameters can also be treated as constants.
12. The argument that Iran doesn't 'need' nuclear power because it is sitting on the world's second largest oil reserves is faulty. Oil is not a direct substitute for electricity – nuclear or otherwise – and it is inefficient to use oil or gas to generate electric power at central stations. The real reason for suspecting that Iran is interested in having at least the potential to produce nuclear weapons is that nuclear power plants in the Middle East were not cost-effective at the time the programme was begun. However with oil selling at $55 or $60 per barrel on the world market, nuclear power becomes much more attractive economically.
13. The leading advocate of this approach has been Amory Lovins. See for instance Lovins et al. (1981); Lovins and Lovins (1987); Fickett et al. (1990); Lovins (1996); and Lovins (1998).
14. The average efficiency of delivered electricity has not increased significantly since 1960. Now, as then, only 33 per cent of the heat content in the fuel burned is delivered to the final user as electric power. Improvements in generators are possible, primarily by introducing so-called 'combined cycle' systems, which means putting a gas turbine at the front end and using the hot exhaust to run a steam turbine. However, combined cycle plants require natural gas or gasified coal to drive the gas turbine. This is not economic in most locations. On the other hand, 'recycling' the by-product energy from both steam turbines and industrial processes, can increase the efficiency of converting fuel to useful work quite dramatically. But these processes typically vent the by-product energy flows to the atmosphere, because the option of co-generating electricity and using it locally has been prohibited by law, to support the monopoly electric utilities.

References

Ahern, J. (1980) *The Exergy Method of Energy System Analysis*, New York: John Wiley.

Ayres, R. U. (1989) *Energy Inefficiency in the US Economy: a New Case for Conservatism*, Laxenburg, Austria: International Institute for Applied Systems Analysis: November. Research Report: RR-89-12.

Ayres, R. U. (1998) 'Eco-thermodynamics: economic and the second law', *Ecological Economics*, 26: 189–209.

Ayres, R. U. and B. Warr (2005) 'Accounting for growth; the role of physical work', *Structural Change and Economic Dynamics*, 16(2): 181–209.

Ayres, R. U., L. W. Ayres and B. Warr (2003) 'Exergy, power and work in the US economy, 1900–1998', *Energy*, 28(3): 219–73.

Ayres, R.U., H. Turton and T. Casten (2007) 'Energy efficiency, sustainability and economic growth', *Energy*, 32(5): 634–48.

Brookes, L. (1990) 'Energy efficiency and economic fallacies', *Energy Policy* 18(3): 199–201.

Casten, T. R. (1998) *Turning Off the Heat: Why America Must Double Energy Efficiency to Save Money and Reduce Global Warming*, Loughton, Essex: Prometheus Press.

Casten, T. and B. Downes (2004) 'Economic growth and the central generation paradigm', paper presented to the International Association of Energy Economists (IAEE) Conference, 17 July.

Fickett, A., G. Gellings, W. Clark and A. Lovins (1990) 'Efficient use of electricity', *Scientific American*, 63(3): 65–74.

Halliday, D. and R. Resnick (1966) *Physics*, New York: Wiley.

Jevons, W. S. (1865) *The Coal Question: an Inquiry Concerning the Progress of the Nation, and the Probable Exhaustion of our Coal Mines*, London: Macmillan.

Kuemmel, R., W. Strassl et al. (1985) 'Technical progress and energy dependent production functions', *Journal of Economics*, 45(3): 285–311.

Lipsey, R. G., K. I. Carlaw and C. T. Bekar (2005) *Economic Transformations: General Purpose Technologies and Long-term Economic Growth*, Oxford: Oxford University Press.

Lovins, A. (1996) 'Negawatts: twelve transitions, eight improvements and one distraction', *Energy Policy*, 24(4): 331–44.

Lovins, A. (1998) 'Energy efficiencies resulting from the adoption of more efficient appliances: another view', *Energy Journal*, 9(2): 155–62.

Lovins, A. and L. H. Lovins (1987) 'Energy: the avoidable oil crisis', *Atlantic Monthly*, 260(6): 22–9.

Lovins, A., L. Lovins, F. Krause and W. Bach (1981) *Least-cost Energy: Solving the CO_2 Problem*, Andover MA: Brickhouse Publication Co.

Mankiw, N. G. (1997) *Macroeconomics*, New York: Worth Publishers.

McKelvey, B. (1968) *The Emergence of Metropolitan America, 1915–1966*, New Jersey: Rutgers University Press.

Mirowski, P. (1989) 'More heat than light: economics as social physics; physics as nature's economics', in C. D. Goodwin (ed.), *Historical Perspectives on Modern Economics*, Cambridge: Cambridge University Press.

Nordhaus, W. D. (1991) 'The cost of slowing climate change: a survey', *Energy Journal*, 12(1): 37–66.

Nordhaus, W. D. (1992) *The DICE Model: Background and Structure of a Dynamic Integrated Model of Climate and the Economy*, New Haven, CT: Cowles Foundation.

Saunders, H. (1992) 'The Khazzoom-Brookes postulate and neoclassical growth', *Energy Journal*, 13(4): 131–48.

Schelling, T. C. (1997) 'The cost of combating global warming', *Foreign Affairs* (November–December): 9.

Strahan, D. (2006) *The Last Oil Shock: a Survival Guide to the Imminent Extinction of Petroleum Man*, London: John Murray Publishers.

7
Exploring Jevons' Paradox

Steve Sorrell

The view that economically justified energy-efficiency improvements will increase rather than reduce energy consumption was first put forward by the British economist, William Stanley Jevons in 1865 (Jevons, 1865). If it were true, 'Jevons' Paradox' would have profound implications for sustainability. It would imply that encouraging energy efficiency as a means of reducing carbon emissions would be futile – rather like a dog chasing its tail. The conventional assumptions of energy analysts, policy-makers, business and lay people alike would be turned on their head and the dominant strategies for achieving sustainability would be undermined. This would be all the more the case if, as seems logical, the Paradox applied to resource efficiency more generally, rather than just energy efficiency. But is the Paradox logically coherent? Do the arguments in its favour stand up to close scrutiny? What empirical evidence is available to suggest that it is correct?

A widely cited formulation of Jevons' Paradox is as follows: 'with fixed real energy prices, energy-efficiency gains will increase energy consumption above what it would be without these gains' (Saunders, 1992). Harry Saunders termed this formulation the *Khazzoom-Brookes (K-B) postulate*, after two contemporary economists (Len Brookes and Daniel Khazzoom) who have been closely associated with the idea. The choice of the term 'postulate' is revealing, since it indicates a starting assumption from which other statements are logically derived and which does not have to be either self-evident or supported by empirical evidence. This interpretation both reflects and encourages a debate that tends to be polarised, theoretical and inconclusive. This chapter seeks to move beyond this by treating the above statement as a hypothesis and exploring some testable implications.

The chapter summarises and critiques the arguments and evidence that have been cited in support of the K-B postulate, focusing in particular on the work of William Stanley Jevons, Len Brookes and Harry Saunders. A notable feature of this work is the lack of reference to quantitative estimates of economy-wide rebound effects. Instead, it comprises a mix of theoretical argument, mathematical modelling, illustrative examples and 'suggestive' evidence from econometric analysis and economic history. It is these 'indirect' sources of evidence that are reviewed in this chapter, together with a number of others that are not cited by the above authors but which appear relevant to their arguments. While most of these sources of evidence make no reference to the rebound effect, they could arguably be used in support of the K-B postulate as well as to challenge conventional wisdom in a number of areas. However, in all cases the evidence is suggestive rather than definitive.

The following section provides a historical context to the debate, including Jevons' nineteenth-century example of the effect of energy-efficiency improvements in steam engines, together with more contemporary examples of comparable 'general-purpose' technologies. This introduces the central theme of this chapter: namely that the arguments and evidence used in support of the K-B postulate are closely linked to a broader question regarding the contribution of energy to productivity improvements and economic growth.

Section two summarises Brookes' arguments in favour of the postulate, identifies some empirical and theoretical weaknesses and examines whether more recent research supports his claims. Section three provides a non-technical summary of Saunders' work, highlighting the dependence of these results on specific theoretical assumptions and the questions this raises for standard economic methodologies. Section four examines some relevant evidence on the contribution of energy to productivity improvements and economic growth and points to the interesting parallels between Brookes' arguments and those of contemporary ecological economists. While the evidence remains ambiguous, the central argument is that energy – and by implication improved energy efficiency – plays a significantly more important role in economic growth than is assumed within mainstream economics. Section five highlights some of the implications of this finding for the economy-wide rebound effect. Section six concludes.

Historical perspectives

Jevons first developed his ideas with reference to coal use and steam engines. His central claim is that:

> it is wholly a confusion of ideas to suppose that the economical use of fuel is equivalent to a diminished consumption. The very contrary is the truth... Every improvement of the engine when effected will only accelerate anew the consumption of coal.
>
> (Jevons, 2001: 99)

He cites the example of the Scottish iron industry, in which:

> the reduction of the consumption of coal, per ton of iron, to less than one third of its former amount, has been followed... by a tenfold increase in total consumption, not to speak of the indirect effect of cheap iron in accelerating other coal consuming branches of industry.
>
> (Jevons, 2001: 103)

According to Jevons, the early Savory engine for pumping floodwater out of coal mines 'consumed no coal because its rate of consumption was too high'. It was only with the subsequent improvements by Watt and others that steam engines became widespread in coal mines, facilitating greater production of lower cost coal which in turn was used by comparable steam engines in a host of applications. One important application was to pump air into blast furnaces, thereby increasing the blast temperatures, reducing the quantity of coal needed to make iron and reducing the cost of iron (Ayres, 2002). Lower cost iron, in turn, reduced the cost of steam engines, creating a positive feedback cycle (Figure 7.1). It also contributed to the development of railways, which lowered the cost of transporting coal and iron, thereby increasing demand for both.

Jevons highlighted the fact that improvements in the thermodynamic efficiency of steam engines were intertwined with broader technical changes, including: 'contrivances, such as the crank, the governor, and the minor mechanism of an engine, necessary for regulating, transmitting, or modifying its power' (Jevons, 2001: 100). These developments were essential to the increased use of steam engines as a source of motive power and demonstrate how energy-efficiency improvements are frequently linked to broader improvements in technology and overall, or 'total factor' productivity.[1]

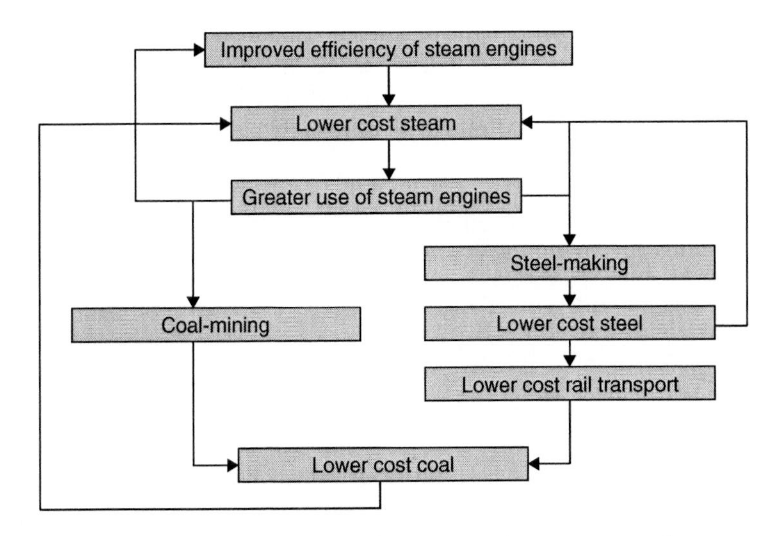

Figure 7.1 Energy efficiency, positive feedbacks and economic growth

Rosenberg (1989) has cited the comparable example of the Bessemer process for steel-making:

> [the Bessemer process] was one of the most fuel saving innovations in the history of metallurgy [and] made it possible to employ steel in a wide variety of uses that were not feasible before Bessemer, bringing with it large increases in demand. As a result, although the process sharply reduced fuel requirements per unit of output, its ultimate effect was to increase . . . the demand for fuel.
>
> (Rosenberg, 1989)

The low-cost Bessemer steel initially found a large market in the production of steel rails, thereby facilitating the growth of the rail industry, and later in a much wider range of applications including automobiles. However, the mild steel produced by the Bessemer process is a very different product to wrought iron (which has a high carbon content) and is suitable for a much wider range of applications. Hence, once again, the improvements in the energy efficiency of production processes are deeply entwined with broader developments in process and product technology. While improved thermodynamic efficiency may form part of – or even a precondition for – such innovations, it does not follow that all the subsequent increase in energy consumption can be attributed to improved energy efficiency.

Table 7.1 Seven centuries of lighting in the UK

Year	*Price of lighting fuel*	*Lighting efficiency*	*Price of lighting services*	*consumption of light per capita*	*Total consumption of light*	*Real GDP per capita*
1300	1.50	0.50	3.0	–	–	0.25
1700	1.50	0.75	2.0	0.17	0.1	0.75
1750	1.65	0.79	2.1	0.22	0.15	0.83
1800	1.0	1	1	1	1	1
1850	0.40	4.4	0.27	3.9	7	1.17
1900	0.26	14.5	0.042	84.7	220	2.9
1950	0.40	340	0.002	1528	5000	3.92
2000	0.18	1000	0.0003	6566	25630	15

Note: 1800 = 1.0 for all indices.
Source: Fouquet and Pearson (2006).

These examples relate to energy-efficiency improvements in the early stages of development of energy-intensive process technologies, producing goods that have the potential for widespread use in multiple applications. It is possible that the same consequences may not follow for energy-efficiency improvements in mature and/or non-energy-intensive process technologies, producing goods that have a relatively narrow range of applications. Similarly, the same consequences may not follow from improvements in consumer technologies that supply energy services with a low own-price elasticity and where energy represents only a small share of total costs (Ayres, 2002).

A historical perspective on rebound effects is provided by Fouquet and Pearson (2006), who present some remarkable data on the price and consumption of lighting services in the UK over a period of seven centuries (Table 7.1). Per capita consumption of lighting services grew much faster than per capita GDP throughout this period, owing in part to continuing reductions in the price per lumen hour. This, in turn, derived from continuing improvements in the energy efficiency of lighting technology, in combination with reductions in the real price of lighting fuel (itself, partly a consequence of improvements in the thermodynamic efficiency of energy supply). In this case, improvements in lighting technology were substantially more important than improvements in energy supply – in the ratio of 180 to 1 over the period 1800–2000.

Per capita lighting consumption increased by a factor of 6566 between 1800 and 2000, largely as a consequence of the falling cost of lighting services relative to income, but also as a result of the boost to per capita GDP provided by the technical improvements in lighting technology. Since lighting efficiency improved by a factor of one thousand, the data suggest that per capita energy consumption for lighting increased by a factor of six. In principle, the direct rebound effect could be determined by estimating the own-price elasticity of lighting services over this period. But this would be a questionable exercise over such a time interval, given the co-evolution and interdependence of the relevant variables. To the extent that the demand for lighting is approaching saturation in many OECD countries, future improvements in lighting efficiency may be associated with smaller rebound effects. Nevertheless, this historical perspective gives cause for concern over the potential of technologies such as compact fluorescents to reduce energy consumption in developing countries.

Energy and economic growth

Time-series data such as that presented in Table 7.1 are difficult to obtain, which partly explains why relatively little research has investigated the causal links between improvements in various measures of energy efficiency and more aggregate measures of economic output and energy consumption. While many studies demonstrate strong correlations between economic output and energy consumption, the extent to which the growth in economic output can be considered a *cause* of the increased energy consumption, or vice versa, remains unclear. It seems likely that there is a synergistic relationship between the two, with each causing the other as part of a positive feedback mechanism (Ayres and Warr, 2002). Hence, to explore Jevons' Paradox further, it seems necessary to investigate the nature, mechanisms and determinants of economic growth – a notoriously complex topic.

The conventional wisdom (as represented by both neoclassical and 'endogenous' growth theory) is that increases in energy inputs play a relatively minor role in economic growth, largely because energy accounts for a relatively small share of total costs (Jones, 1975; Denison, 1985; Gullickson and Harper, 1987; Barro and Sala-I-Martin, 1995). Economic growth is assumed to result instead from the combination of increased capital and labour inputs, changes in the quality of those inputs (e.g. better educated workers) and increases in total factor productivity that are frequently referred to as 'technical change'.

This view has been contested by ecological economists, who argue instead that the increased availability of 'high quality' energy inputs has been the primary driver of economic growth over the last two centuries (Cleveland et al., 1984; Beaudreau, 1998; Kummel et al., 2000).

These authors emphasise that energy carriers differ both in their capacity to perform useful work (captured by the thermodynamic concept of 'exergy') and in their relative economic productivity – reflected by differences in price per kWh (Kaufmann, 1994). So for example, electricity represents a 'higher quality' form of energy than coal. In general, when the 'quality' of energy inputs is accounted for, aggregate measures of energy efficiency are found to be improving more slowly than is commonly supposed (Hong, 1983; Zarnikau, 1999; Cleveland et al., 2000).

Cleveland and his colleagues (1984) claim that a strong link exists between *quality adjusted* energy use and economic output and this link will continue to exist, both temporally and cross-sectionally. This contrasts with the conventional wisdom that energy consumption has been 'decoupled' from economic growth. They also claim that a large component of increased labour productivity over the past 70 years has resulted from empowering workers with increasing quantities of energy, both directly and indirectly as embodied in capital equipment and technology. This contrasts with the conventional wisdom that productivity improvements have resulted from technical change. Other ecological economists argue that the productivity of energy inputs is substantially greater than the share of energy in total costs (Ayres, 2001; Ayres and Warr, 2005) – again in contradiction to the conventional wisdom.

The conventional and ecological perspectives reflect differing assumptions and are supported by conflicting empirical evidence. A difficulty with both is that they confine attention to the relationship between *energy* consumption and economic growth. But the reason that energy is economically significant is that it is used to perform *useful work* – either in the form of mechanical work (including electricity generation) or in the production of heat (Ayres and Warr, 2005). More useful work can be obtained with the same, or less, energy consumption through improved thermodynamic efficiency. Hence, if increases in energy inputs contribute disproportionately to total factor productivity improvements and economic growth, then improvements in thermodynamic efficiency may do the same. Conversely, if increases in energy inputs contribute little to productivity improvements and economic growth, then neither should improvements in thermodynamic efficiency.

The contribution of Len Brookes

Despite their far-reaching implications, Jevons' ideas were neglected until comparatively recently and contemporary advocates of energy efficiency are frequently unaware of them. While a 1980 paper by Daniel Khazzoom stimulated much research and debate on the direct rebound effect (Besen and Johnson, 1982; Einhorn, 1982; Henly et al., 1988; Lovins et al., 1988), most researchers ignored the long-term, macroeconomic implications that were Jevons' primary concern. However, Jevons' arguments have been taken up with some vigour by the British economist, Len Brookes, who has developed coherent arguments in favour of the K-B postulate and combined these with critiques of government energy-efficiency policy (Brookes, 1978, 1984, 1990, 2000, 2004). Brookes' work has prompted a fierce response from critics (Grubb, 1990; Herring and Elliot, 1990; Toke, 1990; Grubb, 1992), to which Brookes has provided a number of responses (Brookes, 1992, 1993).

Brookes (2000: 358) argues that 'The claims of what might be called the Jevons school are susceptible only to suggestive empirical support', since estimating the macroeconomic consequences of individual improvements in energy efficiency is practically impossible. He therefore relies largely on theoretical arguments, supported by indirect sources of evidence, such as historical correlations between various measures of energy efficiency, total factor productivity, economic output and energy consumption (Schurr, 1984, 1985). A key argument runs as follows:

> it has been claimed since the time of Jevons (1865) that the market for a more productive fuel is greater than for less productive fuel, or alternatively that for a resource to find itself in a world of more efficient use is for it to enjoy a reduction in its implicit price with the obvious implications for demand.
>
> (Brookes, 2000: 357)

However, Brookes' use of the term 'implicit price' is confusing. Individual energy-efficiency improvements do not change the price of input energy, but instead lower the effective price of output energy, or useful work. For example, motor-fuel prices may be unchanged following an improvement in vehicle fuel efficiency, but the price per vehicle kilometre is reduced. The 'obvious implications' therefore relate to the demand for useful work, and not to the demand for energy commodities themselves. While the former may be expected to increase, energy demand may either increase or decrease depending upon the price elasticity of demand for useful work and the associated indirect rebound effects.

Of course, the combined impact of multiple energy-efficiency improvements could lower energy demand sufficiently to reduce energy prices and thereby stimulate a corresponding increase in economy-wide energy demand. This forms one component of the economy-wide rebound effect. But while it is obvious that the overall reduction in energy consumption will be less than microeconomic analysis suggests, this theoretical argument appears to be an insufficient basis for claiming that backfire is inevitable.

Brookes also criticises the assumption that the demand for useful work will remain fixed while its marginal cost falls under the influence of raised energy efficiency, and the related assumption that individual energy savings can be added together to produce an estimate of what can be saved over the economy as a whole. In both cases, Brookes is highlighting the persistent neglect of both direct and indirect rebound effects in the conventional assessment of energy-efficiency opportunities. However, arguing that the economy-wide rebound effect is greater than zero is different from arguing that it is greater than one – as the K-B postulate suggests.

Brookes marshals a number of other arguments in support of the K-B postulate that appear more amenable to empirical tests. In doing so, he highlights some important issues regarding the relationship between energy consumption, economic productivity and economic growth. The three most important arguments may be characterised as follows:

- *The 'productivity' argument*: The increased use of higher quality forms of energy (especially electricity) has encouraged technical change, substantially improved total factor productivity and driven economic growth. Despite the substitution of energy for other inputs, this technical change has stimulated a sufficiently rapid growth in economic output that aggregate energy efficiency has *improved* at the same time as aggregate energy consumption has *increased*. Brookes cites two separate, but related sources of empirical evidence in support of this argument. The primary source is the work of Sam Schurr and colleagues on the historical importance of changes in energy quality (notably electrification) in driving US productivity growth (Box 7.1). The second, more indirect source of evidence is the work of Dale Jorgenson and others on the historical direction of technical change. Contrary to standard assumptions, Jorgenson's results suggest that, at the level of individual sectors, technical change has been 'energy-using', meaning that it has increased energy intensity over time rather than reduced it.[2] This work is also cited as suggestive evidence for the K-B postulate by Saunders (1992).
- *The 'accommodation' argument*: Energy-efficiency improvements are claimed to 'accommodate' an energy price shock so that the energy

supply/demand balance is struck at a higher level than if energy efficiency had remained unchanged (Brookes, 1984). While not immediately obvious, this argument appears to rest in part on the assumption that the per capita income elasticity of 'useful' energy demand falls steadily as an economy develops, but is always greater than unity (Brookes, 1972). 'Useful' energy consumption is a quality-adjusted measure of aggregate energy consumption in which different energy types are weighted by their relative economic productivities (Adams and Miovic, 1968).

- *The 'endogeneity' argument*: A common approach to quantifying the 'energy savings' from energy-efficiency improvements is to hold energy intensity fixed at some historic value and estimate what consumption 'would have been' in the absence of those improvements (Geller et al., 2006). The energy savings from energy-efficiency improvements are then taken to be the difference between the actual demand and the counterfactual scenario. But if the energy-efficiency improvements are a *necessary condition* for the growth in economic output, the construction of a counterfactual in this way is misconceived. This argument is not developed in detail by Brookes, but does raise questions over the use of 'decomposition analysis' to explore the rebound effect (Box 7.2).

Box 7.1 Sam Schurr and the rebound effect

Schurr (1960, 1983, 1985) explored trends in US energy consumption, energy productivity and total factor productivity throughout the twentieth century. Energy productivity was defined as the ratio of GDP to total primary energy consumption, with energy being measured on the basis of heat content. Over the period 1920 to 1953, energy, labour and total factor productivity were all found to be growing, while during the period 1953 to 1969, energy productivity was relatively unchanged while total factor productivity continued to grow rapidly. Both periods exhibited falling energy prices relative to other inputs and large increases in energy consumption, and were characterised by a decreasing share of coal in final energy consumption and an increasing share of oil and electricity. Also, in both periods, total factor productivity grew significantly *faster* than energy productivity.

Structural change in the economy and improvements in thermodynamic efficiency provided only a partial explanation of these trends.

Since energy prices were falling in relative terms, energy substituted for other factors of production, thereby reducing energy productivity and improving capital and labour productivity. But these substitution effects were more than outweighed by technological improvements, facilitated by the availability of high-quality energy sources, which greatly improved the overall productive efficiency of the US economy. This meant that economic output increased much faster than energy consumption, owing to the greater productivity of capital and labour. The net result was to produce *falling* energy intensity (as measured by the energy/GDP ratio) alongside *rising* total energy consumption – as the K-B postulate predicts.

Schurr argued that the technological improvements which drove output growth depended crucially upon the increased availability of more 'flexible' forms of energy (oil and electricity) at relatively low costs. These contributed to changes in industrial processes, consumer products and methods of industrial organisation that were quite revolutionary – for example, in transforming the sequence, layout and efficiency of industrial production (Schurr, 1982). Schurr's pioneering contribution, therefore, was to highlight the importance of energy quality for productivity growth.

Brookes uses these observations to support his case for backfire. His argument appears to be that: (a) most improvements in energy productivity are associated with proportionally greater improvements in total factor productivity; (b) improvements in total factor productivity increase economic output, leading to a corresponding increase in demand for inputs; and (c) the resulting increase in demand for energy inputs more than offsets the reduced demand for energy per unit of output. Hence energy consumption increases while aggregate energy intensity falls, as has been observed in most countries over the past century or more.

Sorrell and Dimitropoulos (2007) describe the historical research that forms the basis for these arguments, summarise how Brookes uses this research to support his case and examine in detail whether more recent research confirms or contradicts Brookes' claims. They highlight a number of potential weaknesses, including the following:

- Schurr's work applies primarily to the causal effect of shifts to higher quality fuels (notably electricity), rather than improvements in thermodynamic conversion efficiency or other factors that affect aggregate measures of energy efficiency. The effect of the latter on total factor productivity may not be the same as the effect of the former. Also, the patterns Schurr uncovered may not be as 'normal' as Brookes suggests, since the link between energy productivity and total factor productivity appears to vary greatly, both over time and between different countries and sectors.

- Neither Jorgenson's work itself, nor that of comparable studies, consistently finds technical change to be 'energy-using'. Instead, the empirical results vary widely between different sectors, countries and time periods and are sensitive to minor changes in econometric specification (Norsworthy et al., 1979; Roy et al., 1999; Welsch and Ochsen 2005; Sanstad et al., 2006). Jorgenson's results rest on the erroneous assumption that the rate and direction of technical change is fixed, and more sophisticated models suggest that the magnitude and sign of technical change varies between sectors and types of capital as well as over time (Sue Wing and Eckaus, 2006). Moreover, even if energy-using technical change were to be consistently found, the relationship between this finding and the K-B postulate remains unclear.[3]
- The 'accommodation' argument is based upon a highly simplified theoretical model of the world economy (Brookes, 1984), which is both unconventional in approach and difficult to interpret and calibrate. The model appears to rest in part on the assumption that the income elasticity of 'useful' energy demand declines asymptopically to unity as income increases, thereby allowing economic output to be represented as a linear function of useful energy inputs. While an earlier study by Brookes (1972) provides some support for this hypothesis, this has not been tested by more recent studies of income elasticity of energy demand since these typically measure energy consumption on the basis of heat content (Stern, 2004; Richmond and Kaufmann, 2006).
- The endogeneity argument is rhetorically persuasive but lacks a firm empirical basis. The relative importance of energy-efficiency improvements (however defined) compared to other forms of technical change in encouraging economic growth remains to be established.

In sum, each of these sources of evidence has empirical and theoretical weaknesses and the extent to which they (individually and collectively) support the K-B postulate is open to question. Hence, while Brookes has highlighted some important issues and pointed to sources of evidence that challenge conventional wisdom, he has not provided a convincing case in support of the K-B postulate.

Box 7.2 Endogeneity and the rebound effect

Trends in aggregate quantities may be expressed as the product of a number of variables. For example, economy-wide energy consumption (E) may be expressed into the product of population (P), GDP per capita ($A = Y/P$) and energy use per unit of GDP ($T = E/Y$): $E = PAT$. Decomposition analysis allows the change in energy use over a particular period to be estimated as the sum of the change in each of the right-hand side variables. The 'energy saved' by energy-efficiency improvements over a particular period can then be estimated by comparing current energy consumption with an estimate of what energy consumption 'would have been' had energy intensity (T) remained unchanged. For example, the IEA analysed data from 11 OECD countries over the period 1973 to 1998 to suggest that energy use would have been 50 per cent higher in 1998 if end-use intensity had remained at its 1973 level (Geller et al., 2006).

But this approach is only valid if right-hand side variables are independent of one another – or at least if any dependence is sufficiently small that it can be neglected. In contrast, Brookes argues that improved energy efficiency enables both higher affluence ($A = f(T)$) and higher population ($P = f(T)$): 'it is inconceivable that populations of today could be maintained with the technology of 500 years ago... inanimate energy allied to man's ingenuity is what has permitted the very large increase in output in the last 200 years without which the increase in population would not have occurred. Would this increase (and the associated increase in energy consumption) have occurred if conversion efficiencies had stayed at the abysmally low levels prevailing in the early years of the nineteenth century?'(Brookes, 2000: 359)

To capture this interdependence, the relationship could be better expressed as a system of simultaneous equations (Alcott, 2006):

$$E = f(P, A, T; X_E)$$
$$P = g(E, A, T; X_P)$$
$$A = h(E, P, T; X_A)$$
$$T = i(E, P, A; X_T)$$

Hence, while a reduction in the economy-wide energy/GDP ratio (T) may have a direct effect on energy consumption through the first of these equations, it may also encourage economic growth (*A*), which in turn will increase the total demand for energy (*E*). Over the long term, rising affluence may encourage higher population levels (*P*), which in turn will increase energy consumption (*E*). Each of these changes may in turn influence the energy/GDP ratio (*T*). Hence, a change in one variable could trigger a complex set of adjustments and the final change in energy consumption could be greater or less than the direct change. Under these conditions, decomposition analysis could overestimate the energy savings from improved energy efficiency.

Perhaps the most important insight from Brookes' work is that improvements in energy productivity are frequently associated with proportionally greater improvements in overall or total factor productivity. While Schurr's work provides evidence for this at the level of the national economy, numerous examples from the energy efficiency literature provide comparable evidence at the level of individual sectors and technologies (Pye and McKane, 1998; Worrell et al., 2003; Sorrell et al., 2004). Such examples are frequently used by authors such as Lovins and Lovins (1997) to support the business case for energy efficiency. But if energy-efficient technologies boost total factor productivity and thereby save more than energy costs alone, the argument that rebound effects must be small because the share of energy in total costs is small is undermined. Much the same applies to the contribution of improved energy efficiency to overall productivity improvements and economic growth. But this leaves open the question of whether energy-efficiency improvements (however defined) are necessarily associated with proportionally greater improvements in total factor productivity, or whether (as seems more likely) this is contingent upon particular technologies and circumstances.

The contribution of Harry Saunders

Harry Saunders has shown how the K-B postulate is broadly supported by neoclassical production and growth theory. His work is theoretical and is necessarily based on highly restrictive assumptions. But Saunders does not claim that his work *proves* the K-B postulate; instead, it simply provides suggestive evidence in its favour, given certain *standard* assumptions about how the economy operates.

Saunders (1992) uses neoclassical growth theory to argue that backfire is a likely outcome of 'pure' energy-efficiency improvements – that is, a form of technical change that improves energy productivity while not affecting the productivity of other inputs. In other words, this result does not rely on the contribution of energy to raising capital and labour productivity that is emphasised by Brookes. Neoclassical growth theory also predicts that 'pure' improvements in capital, labour or materials productivity will increase overall energy consumption. Since technical change typically improves the productivity of several inputs simultaneously, these models suggest that most forms of technical change will increase overall energy consumption compared to a scenario in which such improvements are not made.

Saunders' use of the neoclassical growth model was challenged by Howarth (1997), who argued that the failure to distinguish between energy and energy services led to the probability of backfire being overestimated. However, Saunders (2000) subsequently demonstrated that backfire is still predicted by neoclassical theory when an alternative choice is made for the production function used to provide energy services. In a more recent contribution, Saunders (2008) focuses on the potential of different types of production function to generate backfire (see also Chapter 6). Unlike Saunders (1992), this work is also applicable to individual firms and sectors and opens up the possibility of using empirically estimated production functions to estimate the rebound effect from particular technologies in particular sectors (Saunders, 2005).

Saunders (2008) shows how the predicted magnitude of rebound effects depends almost entirely on the choice of the relevant production function – whether at the firm, sector or economy-wide level. Several commonly used production functions are found to be effectively useless in investigating the rebound effect, since the relevant results are the same for whatever values are chosen for key parameters. One popular production function (the constant elasticity of substitution, or CES) is found to be able to simulate rebound effects of different magnitudes, but only if a particular assumption is made about how different inputs are combined. Since this form is widely employed within energy-economic models, Saunders' results raise serious concerns about the ability of such models to accurately simulate rebound effects. An alternative and more flexible functional form (the 'translog') that is widely used in empirical studies is also found to lead to backfire once standard restrictions are imposed on the parameter values to ensure that the behaviour of the function is consistent with economic theory (Saunders, 2008).

There is a substantial empirical literature estimating the parameters of different types of production function at different levels of aggregation and obtaining a good fit with observed data. Hence, if such functions are considered to provide a reasonable representation of real-world economic behaviour, Saunders' work suggests that 'pure' energy-efficiency improvements are likely to lead to backfire. Alternatively, if rebound effects are considered to vary widely in magnitude between different sectors, Saunders' work suggests that standard and widely used economic methodologies cannot be used to simulate them.

The above conclusions apply to pure energy-efficiency improvements. But Saunders (2005) also uses numerical simulations to demonstrate the potential for much larger rebound effects when improvements in energy efficiency are combined with improvements in the productivity of other inputs. Again, if the validity of the theoretical assumptions is accepted, these results suggest that backfire may be a more common outcome than is conventionally assumed.

Saunders' approach is entirely theoretical and therefore severely limited by the assumptions implicit in the relevant models. For instance, technology always comes free, there are only constant returns to scale in production, markets are fully competitive, there is always full employment, qualitative differences in capital and energy are ignored and so on. Indeed, a considerable literature challenges the idea that an 'aggregate' production function for the economy as a whole is a meaningful concept (Fisher, 1993; Temple, 2006) – although this may not necessarily invalidate the use of such functions for representing the behaviour of individual sectors. A particular weakness is the assumption that technical change is costless and autonomous, without explicit representation of the processes that affect its rate and direction. This characteristic limits the capacity of such models to address many policy-relevant questions. More recent developments in so-called 'endogenous growth theory' have overcome this weakness to a large extent, but to date no authors have used such models to explore the rebound effect. However, since what is at issue is the consequences of energy-efficiency improvements, the source of those improvements is arguably a secondary concern.

Overall, Saunders' work suggests that significant rebound effects can exist in theory, backfire is quite likely and this result is robust to different model assumptions. Since these results derive from a contested theoretical framework, they are suggestive rather than definitive. But they deserve to be taken seriously.

Energy productivity and ecological economics

In his 1984 paper, Brookes quotes Sam Schurr's observation that 'it is energy that drives modern economic systems rather than such systems creating a demand for energy' (Brookes, 1984: 383). This highlights an underlying theme in much of Brookes' work: namely that energy plays a more important role in driving productivity improvements and economic growth than is conventionally assumed. But precisely the same claim is made by ecological economists such as Cleveland and his colleagues (1984), who attribute a large component of the productivity increases over the past century to the increasing availability of high-quality energy sources. This leads them to express scepticism over the scope for decoupling economic growth from increased energy consumption.

Ecological economists have not directly investigated the rebound effect, but their work arguably provides suggestive support for the K-B postulate in much the same way as Schurr's research on the historical determinants of US productivity growth. Four examples of this work are briefly described below.

First, analysis by Kaufmann (1992, 2004) and others suggests that historical reductions in energy/GDP ratios owe much more to structural change and shifts towards 'high-quality' fuels than to technological improvements in energy efficiency. By neglecting changes in energy quality, conventional analysts may have come to incorrect conclusions regarding the rate and direction of technical change and its contribution to reduced energy consumption. Kaufmann (1992) suggests that, not only does the energy/GDP ratio reflect the influence of factors *other* than energy-saving technical change, but these other factors may be *sufficient* to explain the observed trends. Hence, the observed improvements in the thermodynamic efficiency of individual devices at the micro level do not appear to have significantly contributed to the observed reduction in energy intensity at the macro level. As with the work of Jorgenson and others, this suggests that the conventional assumptions of energy-economic models may be flawed.

Second, both neoclassical and ecological economists have used modern econometric techniques to test the direction of causality between energy consumption and GDP (Stern, 1993; Chontanawat et al., 2006; Lee, 2006; Yoo, 2006; Zachariadis, 2006). The assumption is that if GDP growth is the cause of increased energy consumption then a change in the GDP growth rate should be followed by a change in energy consumption and vice versa. It is argued that if causality runs from GDP to energy consumption then energy consumption may be reduced without adverse effects on economic growth, while if causality runs the other way round a reduction in energy use may negatively affect economic growth. While the results of such studies are frequently contradictory, most of them neglect changes in energy quality. When energy quality is taken into account, the causality appears to run from energy consumption to GDP – as ecological economists suggest (Stern, 1993, 2000).

Third, historical experience provides very little support for the claim that increases in income will lead to declining energy consumption (Stern, 2004; Richmond and Kaufmann, 2006). While the income elasticity of aggregate energy consumption may be both declining and less than one in OECD countries, there is no evidence that it is negative (or is soon to become negative). Again, neglect of changes in fuel mix and energy prices may have led earlier studies to draw misleading conclusions regarding the extent to which energy consumption has been decoupled from GDP (Kaufmann, 2004).

Finally, ecological economists have developed a number of alternatives to the conventional models of economic growth (Kummel et al., 1985; Beaudreau, 1998; Kummel et al., 2000; Ayres and Warr, 2005). A key feature of these models is a departure from the traditional assumption that the productivity of each input is proportional to the share of that input in the value of output. Instead, the productivity of each input is estimated directly from a production function. These models are found to reproduce historical trends in economic growth extremely well, without attributing any role to technical change. This is in contrast to conventional theories of economic growth, which attribute much of the increase in output to technical change.[4] The marginal productivity of energy inputs is found to be around ten times larger than their cost share, implying that improvements in energy productivity could have a dramatic effect on economic growth and therefore on economy-wide energy consumption – in other words, the rebound effect could be very large.

Of particular interest is the work by Ayres and Warr (2005), who combine historical data on the 'exergy' content of fuel inputs and second-law thermodynamic conversion efficiencies to develop a unique time series of the exergy output of conversion devices (termed useful work) in the US economy over the past century (Table 7.2). They show that useful work inputs to the US economy have grown by a factor of 18 over the past 100 years, implying that the useful work obtained from fuel resources has grown much faster than the consumption of fuels themselves, owing to substantial improvements in thermodynamic

Table 7.2 Trends in second-law conversion efficiencies of primary conversion processes in the US (average % efficiency in specified year)

Year	*Electricity generation and distribution*	*Transportation*	*High temperature process heat (steel)*	*Medium temperature process heat (steam)*	*Low temperature space heat*
1900	3.8	3.0	7	5	0.25
1970	32.5	8.0	20	14	2
1990	33.3	13.9	25	20	3

Source: Ayres et al. (2003).

conversion efficiencies. By including useful work in their production function, rather than primary energy, Ayres and Warr (2005) obtain an extremely good fit to US GDP trends over the past century, thereby eliminating the need for a multiplier for technical change. The implication is that improvements in thermodynamic conversion efficiency provide a quantifiable surrogate for all forms of technical change that contribute to economic growth. Far from being a minor contributor to economic growth, improvements in thermodynamic efficiency become the dominant driver – obviating the need for alternative measures of technological change.

The implication of this work is that energy is more productive than is suggested by its small share of total costs. This is precisely the argument that Schurr made and which appears to underlie some of Brookes' arguments in favour of backfire. However, the empirical evidence in support of this perspective remains patchy. For example, the results of econometric investigations of causality relationships between energy and GDP remain ambiguous and the policy implications that are drawn are frequently oversimplified (Zachariadis, 2006). Also, the statistical form of causality that is being measured here (so-called 'Granger causality') is not the same as causality as conventionally understood and conventional notions of causality may be problematic for systems as complex as modern economies. In a similar manner, the different variants of 'ecological growth models' rely upon an unusual and oddly behaved production function, provide results that are difficult to reconcile with each other and appear vulnerable to bias from a number of sources that could potentially invalidate the results (Sorrell and Dimitropoulos, 2007). As a result, claims that the marginal productivity of energy is an order of magnitude larger than its cost share, or that improvements in thermodynamic conversion efficiency can act as a suitable proxy for technical change, must be treated with considerable caution.

Unfortunately, the different assumptions of conventional and ecological perspectives on economic growth seem to have prevented an objective comparison of their methods and conclusions. Convincing evidence of the disproportionate contribution of energy to productivity improvements and economic growth therefore remains elusive. Moreover, even if this were to be accepted, the link from this evidence to the K-B postulate remains ambiguous and indirect.

The neoclassical assumption appears to be that capital, labour and energy inputs have *independent* and *additive* effects on economic output, with any residual increase being attributed to exogenous technical change. Endogenous growth theory has modified these assumptions, but still attributes a relatively minor role to energy. In contrast, the ecological assumption appears to be that capital, labour and energy are *interdependent* inputs that have *synergistic* and *multiplicative* effects on economic output, and that the increased availability of low-cost, high-quality energy sources has provided a necessary condition for most historical improvements in economic productivity. A bridge between the two could potentially be provided by Toman and Jemelkova's (2003) observation that increased inputs of useful work (or energy services) may *enhance* the productivity of capital and labour:

> when the supply of energy services is increased, there is not just more energy to be used by each skilled worker or machine; the productivity with which every unit of energy is used also rises. If all inputs to final production are increased in some proportion, final output would grow in greater proportion because of the effect on non-energy inputs.
>
> (Toman and Jemelkova, 2003: 100)

Schurr's (1983, 1984, 1985) account of the impact of electrification (and especially electric motors) on the organisation and productivity of US manufacturing provides an example of this process. But the extent to which such patterns have applied in other sectors and time periods needs to be determined empirically. If such a situation is the norm, the increased availability of high-quality energy may be a primary driver of economic activity. But if the increased availability of high-quality energy inputs has a disproportionate impact on productivity and economic growth, then improvements in thermodynamic conversion efficiency may do the same, because both increase the useful work available from conversion devices. Under these conditions, the rebound effect could be large and potentially greater than unity.

Implications

The K-B postulate implies that *all* economically justified energy-efficiency improvements will increase energy consumption above where it would be without those improvements. Since this is a counterintuitive claim for many people, it requires strong supporting evidence if it is to gain widespread acceptance. The main conclusion from this chapter is that such evidence does not yet exist. The theoretical and empirical evidence cited in favour of the postulate contains a number of weaknesses and inconsistencies and most is only indirectly relevant to the rebound effect. Nevertheless, the arguments and evidence deserve more serious attention than they have received to date. Much of the evidence points to economy-wide rebound effects being larger than is conventionally assumed and to energy playing a more important role in driving productivity improvements and economic growth than is conventionally assumed.

The possibility of large economy-wide rebound effects has been dismissed by a number of leading energy analysts (Howarth, 1997; Lovins, 1998; Laitner, 2000; Schipper and Grubb, 2000). But it becomes more plausible *if* it is accepted that energy-efficiency improvements are frequently associated with improvements in the productivity of other inputs. If this is the case, then rebound effects need not necessarily be small just because the share of energy in total costs is small. Future research should therefore investigate the extent to which improvements in energy efficiency (however defined and measured) are associated with broader improvements in economic productivity, and the circumstances under which economy-wide rebound effects are more or less likely to be large (Figure 7.2).

Rebound effects may be particularly large for the energy-efficiency improvements associated with so-called 'general-purpose technologies', such as steam engines and computers. General-purpose technologies (GPTs) have a wide scope for improvement and elaboration, are applicable across a broad range of uses, have potential for use in a wide variety of products and processes and have strong complementarities with existing or potential new technologies (Lipsey et al., 2005). Steam engines provide a paradigmatic illustration of a GPT in the nineteenth century, while electric motors provide a comparable illustration for the early twentieth century. The former was used by Jevons to support the case for backfire,

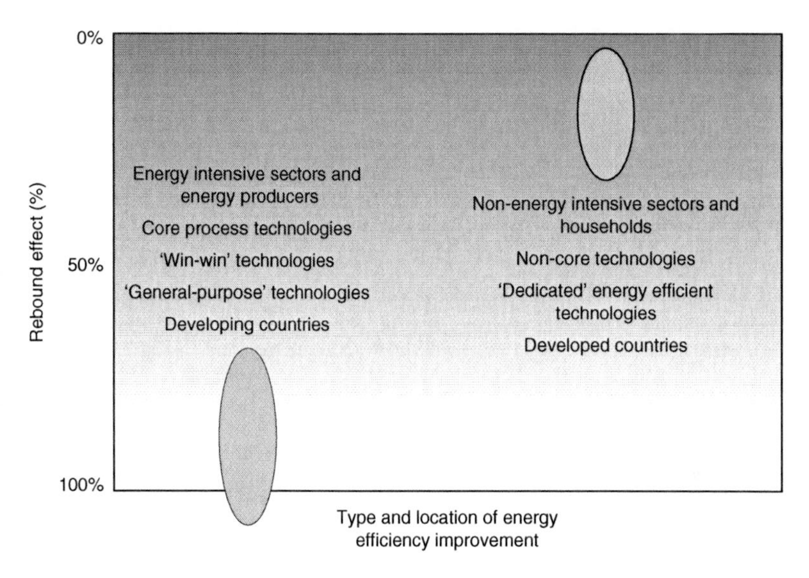

Figure 7.2 Conditions under which rebound effects may be large or small

while the latter formed a key component of Schurr's work which was subsequently cited by Brookes as suggestive evidence for backfire.

The key to unpacking the K-B postulate may therefore be to distinguish the energy-efficiency improvements associated with GPTs from other forms of energy-efficiency improvement. The K-B postulate seems more likely to hold for the former, particularly when these are used by producers and when the energy-efficiency improvements occur at an early stage of development and diffusion of the technology. The opportunities offered by these technologies have such long-term and significant effects on innovation, productivity and economic growth that economy-wide energy consumption is increased. In contrast, the K-B postulate seems less likely to hold for dedicated energy-efficiency technologies such as improved thermal insulation, particularly when these are used by consumers or when they play a subsidiary role in economic production. These technologies have smaller effects on productivity and economic growth, with the result that economy-wide energy consumption may be reduced.

The implication is that energy policy should focus on encouraging dedicated energy-efficiency technologies, rather than improving the energy efficiency of GPTs. However, these categories are poorly defined and the boundaries between them are blurred. Moreover, even if GPTs can meaningfully be distinguished from other forms of technology, continued economic growth is likely to depend upon the diffusion of new types of GPT that may increase aggregate energy consumption.

Summary

The case for the K-B postulate is not based upon empirical estimates of rebound effects. Instead, it relies largely upon theoretical arguments, backed up by empirical evidence that is both suggestive and indirect. Disputes over the postulate therefore hinge in part on competing theoretical assumptions. While historical experience demonstrates that substantial improvements in energy efficiency have occurred alongside increases in economic output, total factor productivity and overall energy consumption, this does not provide sufficient evidence for the K-B postulate since the causal links between these trends remain unclear.

This chapter has reviewed the strengths and weaknesses of the arguments and evidence in favour of backfire presented by Len Brookes and Harry Saunders. While neither author provides a totally convincing case in favour of the K-B postulate, their work poses an important challenge to conventional wisdom. Of particular interest is the apparent similarity between some of Brookes' arguments and the heterodox claim that the increased availability and quality of energy inputs is the primary driver of economic activity. But energy is only economically productive because it provides useful work. Hence, if increases in energy inputs contribute disproportionately to economic growth, then improvements in thermodynamic efficiency should do the same, since both provide more useful work. The dispute over the K-B postulate may therefore be linked to a broader question of the contribution of energy to economic growth.

The debate over the K-B postulate would benefit from further distinctions between different types of energy-efficiency improvement. In particular, the K-B postulate seems more likely to hold for energy-efficiency improvements associated with the early stage of diffusion of 'general-purpose technologies', such as electric motors in the early twentieth century. It may be less likely to hold for the later stages of diffusion of these technologies, or for 'dedicated' energy-efficiency technologies such as improved thermal insulation. However, these categories are poorly defined and the boundaries between them are blurred. Overall, while it seems unlikely that all energy-efficiency improvements will lead to backfire, we still have much to learn about the factors that make backfire more or less likely to occur.

Acknowledgements

This chapter is based upon a comprehensive review of the evidence for rebound effects, conducted by the UK Energy Research Centre (Sorrell, 2007). An earlier version of this chapter is contained in Sorrell (2007), while the research upon which it is based is reported in detail in Sorrell and Dimitropoulos (2007). These reports are available to download from the UKERC website. The financial support of the UK Research Councils is gratefully acknowledged.

Considerable thanks are due to John Dimitropoulos (Sussex Energy Group) for his contribution to this work and also to Harry Saunders for his extremely helpful comments and encouragement. The author is also very grateful for the advice and comments received from Manuel Frondel, Karsten Neuhoff, Jake Chapman, Nick Eyre, Serban Scrieciu, Blake Alcott, Len Brookes, John Feather, Gordon Mackerron, Horace Herring, Lester Hunt and Paolo Agnolucci, Jim Skea, Rob Gross and Phil Heptonstall. The usual disclaimers apply.

Notes

1. The economic productivity (or efficiency) of a 'factor' input such as energy is given by the ratio of output to input for that factor. Total factor productivity (TFP) is normally defined as the rate of growth of economic output minus the weighted sum of the rate of growth of inputs – with each input being weighted by its share in the value of output (Sorrell and Dimitropoulos, 2007). Unlike changes in individual factor productivity, improvements in total factor productivity are always desirable, since they indicate that more output is being obtained from the same quantity of inputs. Standard 'growth accounting' techniques estimate total factor productivity as the residual growth in output that is not explained by the growth of inputs. Such improvements are frequently attributed to 'technical change'.
2. Jorgenson and Fraumeni (1981) used econometric techniques to investigate the impact of the energy price rises of the 1970s on US manufacturing productivity. They employed the conventional neoclassical distinction between price-induced substitution between factor inputs and autonomous (i.e. non-price induced) technical change. They estimated the rate of change in the share of energy costs in the value of output of US manufacturing sectors (holding input prices constant) and found that, in 29 out of 35 sectors, the share of energy costs increased over time. This is termed 'energy-using' technical change. Generally, we would expect energy-using technical change to be associated with increasing energy intensity. However, this may not always be the case, since it also depends upon the rate of change in total factor productivity (Sanstad et al., 2006). With energy-using technical change an increase in the price of energy will lower total factor productivity.

3. The relevance of this work to the rebound effect is not made clear by either Brookes or Saunders. The primary implication is that technical change has frequently reduced energy efficiency and thereby increased overall energy consumption, even while other factors (such as structural change) have acted to decrease energy consumption. Not only is this the opposite to what is conventionally assumed in energy-economic models, it is also opposite to what is required for an empirical estimate of the rebound effect. At the same time, technical change has clearly improved the thermodynamic conversion efficiency of individual devices, such as motors and boilers. What Jorgenson and Fraumeni's work suggests, therefore, is that these improvements in thermodynamic efficiency have not necessarily translated into improvements in more aggregate measures of energy intensity at the level of industrial sectors. Similarly, a more recent study by Sue Wing and Eckhaus (2006) suggests that this has not necessarily translated into improvements in more aggregate measures of energy intensity for particular types of capital (e.g. machinery). In other words, improvements in energy efficiency at one level of aggregation may have contributed to greater energy consumption at a higher level of aggregation. Hence, the relevance of these results may hinge in part upon the appropriate choice of independent variable for the rebound effect.
4. Traditional neoclassical growth models estimate 'technical change' as the residual growth in output that is not explained by the growth of inputs. Early growth models attributed as much as 70 per cent of the growth in output to technical change, but later studies have shown how the proportion of growth that is attributed to technical change depends upon how the inputs are measured (Jorgenson and Griliches 1967). In neoclassical growth models, technical change is typically represented by a simple time trend. Modern theories of economic growth seek to make the source and direction of technical change endogenous.

References

Adams, F. G. and P. Miovic (1968) 'On relative fuel efficiency and the output elasticity of energy consumption in Western Europe', *Journal of Industrial Economics*, 17(1): 41–56.

Alcott, B. (2006) 'Assessing energy policy: should rebound count?' MPhil thesis, Department of Land Economy, University of Cambridge, Cambridge.

Ayres, R. U. (2001) 'The minimum complexity of endogenous growth models: the role of physical resource flows', *Energy*, 26(9): 817–38.

Ayres, R. U. (2002) *Resources, Scarcity, Technology and Growth*, INSEAD Working Paper No. 2002/118/EPS/CMER, Centre for the Management of Environmental Resources, INSEAD, Fontainebleau.

Ayres, R. U. and B. Warr (2002) *Two Paradigms of Production and Growth*, CMER Working Paper, INSEAD, Fontainbleau, France.

Ayres, R. U. and B. Warr (2005) 'Accounting for growth: the role of physical work', *Structural Change and Economic Dynamics*, 16(2): 181–209.

Ayres, R. U., L. W. Ayres and B. Warr (2003) 'Exergy, power and work in the US economy, 1900–1998', *Energy*, 28(3): 219–73.

Barro, R. J. and X. Sala-I-Martin (1995) *Economic Growth*, New York: McGraw-Hill.

Beaudreau, B. C. (1998) *Energy and Organisation: Group and Distribution Re-examined*, Westport, CT: Greenwood Press.

Besen, S. M. and L. L. Johnson (1982) 'Comment on "Economic implications of mandated efficiency standards for household appliances"', *Energy Journal*, 3(1): 110–16.

Brookes, L. G. (1972) 'More on the output elasticity of energy consumption', *Journal of Industrial Economics*, 21(1): 83–92.

Brookes, L. G. (1978) 'Energy policy, the energy price fallacy and the role of nuclear energy in the UK', *Energy Policy*, 6(2): 94–106.

Brookes, L. G. (1984) 'Long-term equilibrium effects of constraints in energy supply', in L. G. Brookes and H. Motamen (eds), *The Economics of Nuclear Energy*, London: Chapman and Hall.

Brookes, L. G. (1990) 'The greenhouse effect: the fallacies in the energy efficiency solution', *Energy Policy*, 18(2): 199–201.

Brookes, L. G. (1992) 'Energy efficiency and economic fallacies – a reply', *Energy Policy*, 20(5): 390–2.

Brookes, L. G. (1993) 'Energy efficiency fallacies – the debate concluded', *Energy Policy*, 21(4): 346–7.

Brookes, L. G. (2000) 'Energy efficiency fallacies revisited', *Energy Policy*, 28(6–7): 355–66.

Brookes, L. G. (2004) 'Energy efficiency fallacies – a postscript', *Energy Policy*, 32(8): 945–7.

Chontanawat, J., L. Hunt and R. Pierse (2006) *Causality between Energy Consumption and GDP: Evidence from 30 OECD and 78 non-OECD countries*, Working Paper, Department of Economics, University of Surrey, Guildford.

Cleveland, C. J., R. Costanza, C. A. S. Hall and R. K. Kaufmann (1984) 'Energy and the US economy: a biophysical perspective', *Science*, 225: 890–7.

Cleveland, C. J., R. K. Kaufmann and D. I. Stern (2000) 'Aggregation and the role of energy in the economy', *Ecological Economics*, 32: 301–17.

Denison, E. F. (1985) *Trends in American Economic Growth 1929–1982*, Washington, DC: Brookings Institute.

Einhorn, M. (1982) 'Economic implications of mandated efficiency standards for household appliances: an extension', *Energy Journal*, 3(1): 103–9.

Fisher, F. M. (1993) *Aggregation: Aggregate Production Functions and Related Topics*, Cambridge, MA: MIT Press.

Fouquet, R. and P. Pearson (2006) 'Seven centuries of energy services: the price and use of light in the United Kingdom (1300–1700)', *Energy Journal*, 27(1): 139–77.

Geller, H., P. Harrington, A. H. Rosenfeld, S. Tanishima and F. Unander (2006) 'Policies for increasing energy efficiency: thirty years of experience in OECD countries', *Energy Policy*, 34: 556–73.

Grubb, M. J. (1990) 'Energy efficiency and economic fallacies', *Energy Policy*, 18(8): 783–5.

Grubb, M. J. (1992) 'Reply to Brookes', *Energy Policy*, 20(5): 392–3.

Gullickson, W. and M. J. Harper (1987) 'Multi factor productivity in US manufacturing 1949–1983', *Monthly Labour Review* (October): 18–28.

Henly, J., H. Ruderman and M. D. Levine (1988) 'Energy savings resulting from the adoption of more efficient appliances: a follow-up', *Energy Journal*, 9(2): 163–70.

Herring, H. and M. Elliot (1990) 'Letter to the editor', *Energy Policy*, 18(8): 786.

Hong, N. V. (1983) 'Two measures of aggregate energy production elasticities', *Energy Journal*, 4(2): 172–7.

Howarth, R. B. (1997) 'Energy efficiency and economic growth', *Contemporary Economic Policy*, 15(4): 1.

Jevons, W. S. (2001) 'Of the economy of fuel', *Organization & Environment*, 14(1): 99–104.

Jones, H. G. (1975) *An Introduction to Modern Theories of Economic Growth*, London: Van Nostrand Reinhold.

Jorgenson, D. and B. Fraumeni (1981) 'Relative prices and technical change', in E. R. Berndt and B. C. Field (eds), *Modeling and Measuring Natural Resource Substitution*, Boston: MIT Press.

Jorgenson, D. W. and Z. Griliches (1967) 'The explanation of productivity change', *Review of Economic Studies*, 34(3): 249–82.

Kaufmann, R. K. (1992) 'A biophysical analysis of the energy/real GDP ratio: implications for substitution and technical change', *Ecological Economics*, 6(1): 35–56.

Kaufmann, R. K. (1994) 'The relation between marginal product and price in US energy markets: implications for climate change policy', *Energy Economics*, 16(2): 145–58.

Kaufmann, R. K. (2004) 'The mechanisms of autonomous energy efficiency increases: a cointegration analysis of the US energy/GDP ratio', *Energy Journal*, 25(1): 63–86.

Kummel, R., D. Lindenberger and W. Eichhorn (2000) 'The productive power of energy and economic evolution', *Indian Journal of Applied Economics*, 8: 231–62.

Kummel, R., W. Strassl, A. Gossner and W. Eichhorn (1985) 'Technical progress and energy dependent production functions', *Nationalökonomie – Journal of Economics*, 3: 285–311.

Laitner, J. A. (2000) 'Energy efficiency: rebounding to a sound analytical perspective', *Energy Policy*, 28(6–7): 471–5.

Lee, C. C. (2006) 'The causality relationship between energy consumption and GDP in G-11 countries revisited', *Energy Policy*, 34(9): 1086–93.

Lipsey, R. G., K. I. Carlaw and C. T. Bekar (2005) *Economic Transformations: General Purpose Technologies and Long-term Economic Growth*, Oxford: Oxford University Press.

Lovins, A. B. (1998) 'Further comments on red herrings', letter to the *New Scientist*, No. 2152, 18 September.

Lovins, A. B. and L. H. Lovins (1997) *Climate: Making Sense and Making Money*, Old Snowmass, Colorado: Rocky Mountain Institute.

Lovins, A. B., J. Henly, H. Ruderman and M. D. Levine (1988) 'Energy saving resulting from the adoption of more efficient appliances: another view; a follow-up', *Energy Journal*, 9(2): 155.

Norsworthy, J. R., M. J. Harper and K. Kunz (1979) 'The slowdown in productivity growth: an analysis of some contributing factors', *Brookings Papers in Economic Activity*, 2: 387–41.

Pye, M. and A. McKane (1998) 'Enhancing shareholder value: making a more compelling energy efficiency case to industry by quantifying non-energy benefits', Proceedings of the 1999 Summer Study on Energy Efficiency in Industry, Washington, DC.

Richmond, A. K. and R. K. Kaufmann (2006) 'Energy prices and turning points: the relationship between income and energy use/carbon emissions', *Energy Journal*, 27(4): 157–79.

Rosenberg, N. (1989) 'Energy efficient technologies: past, present and future perspectives', paper presented at the conference How Far Can the World Get on Energy Efficiency Alone?, ORNL, August.

Roy, J., J. Sathaye, A. H. Sanstad, P. Mongia and K. Schumacher (1999) 'Productivity trends in India's energy intensive industries', *Energy Journal*, 20(3): 33–61.

Sanstad, A. H., J. Roy and S. Sathaye (2006) 'Estimating energy-augmenting technological change in developing country industries', *Energy Economics*, 28: 720–9.

Saunders, H. D. (1992) 'The Khazzoom-Brookes postulate and neoclassical growth', *Energy Journal*, 13(4): 131.

Saunders, H. D. (2000) 'Does predicted rebound depend on distinguishing between energy and energy services?' *Energy Policy*, 28(6–7): 497–500.

Saunders, H. D. (2005) 'A calculator for energy consumption changes arising from new technologies', *Topics in Economic Analysis and Policy*, 5(1): 1–31.

Saunders, H. D. (2008) 'Fuel conserving (and using) production functions', *Energy Economics*, in press.

Schipper, L. and M. Grubb (2000) 'On the rebound? Feedback between energy intensities and energy uses in IEA countries', *Energy Policy*, 28(6–7): 367–88.

Schurr, S. (1982) 'Energy efficiency and productive efficiency: some thoughts based on the American experience', *Energy Journal*, 3(3): 3–14.

Schurr, S. H. (1983) 'Energy abundance and economic progress', *Annual Review of Energy*, 10(3–4): 111–17.

Schurr, S. H. (1984) 'Energy, economic growth and human welfare', in L. G. Brookes and H. Motamen (eds), *The Economics of Nuclear Energy*, London: Chapman and Hall.

Schurr, S. H. (1985) 'Energy conservation and productivity growth: can we have both?' *Energy Policy*, 13(2): 126–32.

Schurr, S. H., B. C. Netschert, V. E. Eliasberg, J. Lerner and H. H. Landsberg (1960) *Energy in the American Economy*, Baltimore: Resources for the Future and Johns Hopkins University Press.

Sorrell, S. (2007) *The Rebound Effect: an Assessment of the Evidence for Economy-wide Energy Savings from Improved Energy Efficiency*, London: UK Energy Research Centre.

Sorrell, S. and J. Dimitropoulos (2007) *UKERC Review of Evidence for the Rebound Effect: Technical Report 5: Energy, Productivity and Economic Growth Studies*, London: UK Energy Research Centre.

Sorrell, S., E. O'Malley, J. Schleich and S. Scott (2004) *The Economics of Energy Efficiency: Barriers to Cost-Effective Investment*, Cheltenham: Edward Elgar.

Stern, D. I. (1993) 'Energy and economic growth in the USA: a multivariate approach', *Energy Economics*, 15(2): 137–50.

Stern, D. I. (2000) 'A multivariate cointegration analysis of the role of energy in the US macroeconomy', *Energy Economics*, 22: 267–83.

Stern, D. I. (2004) 'The rise and fall of the environmental Kuznets curve', *World Development*, 32(8): 1419–39.

Sue Wing, I. and R. S. Eckaus (2006) *The Decline in US Energy Intensity: Its Origins and Implications for Long-Run CO_2 Emission Projections*, Working Paper, Kennedy School of Government, Harvard University, Boston.

Temple, J. (2006) 'Aggregate production functions and growth economics', *International Review of Applied Economics*, 20(3): 301–17.

Toke, D. (1990) 'Increasing energy supply not inevitable', *Energy Policy*, 18(7): 671–3.

Toman, M. and B. Jemelkova (2003) 'Energy and economic development: an assessment of the state of knowledge', *Energy Journal*, 24(4): 93–112.

Welsch, H. and C. Ochsen (2005) 'The determinants of aggregate energy use in West Germany: factor substitution, technological change, and trade', *Energy Economics*, 27: 93–111.

Worrell, J. A., M. Ruth, H. E. Finman and J. A. Laitner (2003) 'Productivity benefits of industrial energy efficiency measures', *Energy*, 1081–98.

Yoo, S. H. (2006) 'The causal relationship between electricity consumption and economic growth in the ASEAN countries', *Energy Policy*, 34(18): 3573–82.

Zachariadis, T. (2006) *On the Exploration of Causal Relationships between Energy and the Economy*, Discussion Paper 2006-05, Department of Economics, University of Cyprus.

Zarnikau, J. (1999) 'Will tomorrow's energy efficiency indices prove useful in economic studies?' *Energy Journal*, 20(3): 139–45.

Part III

Rebound Effects and Sustainable Consumption

8

Time-use Rebound Effects: an Activity-based View of Consumption

Mikko Jalas

Energy-efficiency innovations frequently contribute to and depend on changes in the time allocation of households. In this chapter I argue that a focus on such changes in time use can broaden the scope of the rebound debate and help to account for transformative types of rebound effects. In simple terms, I argue that one can reframe the debate by shifting the focus from monetary effects to the time use of individual consumers.

The suggested time use approach translates rather readily into the rebound discussion. One can distinguish price effects in which efficiency innovations, i.e. time-saving improvements, increase the demand for the same product or service. On the other hand, such innovations also cause reallocation of time and hence structural effects in the time budget. However, what is distinctive with the time use approach is that it better captures what Greening et al. (2000) call transformative rebound effects. These are effects that are not only structural, but contribute to economic expansion through, for example, increased labour participation.

Consider, for example, the popular claim that outsourcing domestic activities is an effective organisational innovation to improve energy efficiency (see e.g. von Weizsäcker et al., 1997; Meijkamp, 1998; Goedkoop et al., 1999; Mont, 2002; Hirschl et al., 2003; Halme et al., 2004). It is argued that a market actor delivering services such as lawn mowing will offer energy-efficiency benefits compared to the ownership and use of private lawn mowers. If true, such innovations appear highly desirable since they may at the same time create structural changes towards services and support labour policy objectives. However, such efficiency improvements also impact the time budget of the individual citizens and encourage increased hours of paid work thereby pushing transformative changes in the economy. Such changes thus run counter to the sufficiency-minded arguments for *reducing* the hours of work (e.g.

Schor 1991, 2005; Sanne, 2000; Reisch, 2001; Sanches, 2005; see also Chapter 11). More generally, as the outsourcing of domestic activities is gaining in popularity as a policy tool to encourage sustainable consumption, it is relevant to theorise the wider impacts of such potential and ongoing shifts.

Previously, Schipper and colleagues (1989) have considered the relationship between time use and energy consumption. Their analysis brought forward this relation as well as the observation that the transfer of home-based activities to more distant locations increases transportation demand. Binswanger (2001) has addressed the secondary impacts of time saving innovations in consumption. His claim is that such innovations are frequently energy intensive and prone to occur when the price of energy is low relative to that of labour. However, Binswanger does not highlight the transformational effects discussed below. To that end, the focus of this chapter is the relationship between outsourcing of domestic activities, market expansion and increased energy demand. Most crucially, however, this chapter attempts to understand the saturation of demand by suggesting that time can be viewed not as a scarce, money-like resource, but rather as a lived experience and an 'output' of economic activities. This is to assume that much of modern consumption is not goal-directed activity, but informed by the benefits and meanings of the activity itself. This argument challenges the conventional economic outlook, including the notion of scarcity as the underlying premise of the rebound debate.

The suggested time use approach will lead to less analytical rigour and impede the quantification of the rebound effects. On the other hand, by highlighting some important mechanisms of market expansion, it is possible to broaden the scope of the rebound debate. Furthermore, while the quantification of time use rebound effects may appear as a distant goal for research, the suggested approach does not exclude empirical examination. Time use data exist and avail themselves for analysis of changes in time allocation and the energy intensity of non-work activities. I will first elaborate on the time use approach in general and then consider the particular implications of this approach for the debate on rebound effects. Thereafter I will consider empirical methods and available data sets.

A time use perspective on consumption

The notion of a time use rebound effect is built on a 'temporalised' view of human activity according to which consumption consists of a set of activities in which humans combine their own time with market-provided goods and services. Hence, this approach follows the tradition of household economics (e.g. Becker, 1965; Lancaster, 1966). The purpose of this section is to introduce a selective set of concepts and ideas which stem from this tradition and which I have applied in my earlier writings on energy efficiency and rebound effects in everyday life (Jalas 2002, 2005a, 2006). Yet, my relation to this tradition is uneasy at best; contradicting the 'Beckerian' approach I emphasise the intrinsic meanings of human activity. This is to argue that much of modern consumption cannot be thought of as goal-directed economic activity, and that process-benefits and process-utility outweigh instrumental goal-oriented reasoning in everyday life.

Following this line of reasoning and departing from traditional household economics, I have previously argued that 'consumption should be regarded as a set of temporal activities in which consumers utilize or engage with the various products of the industrial systems and through which the resource flows pass, virtually or in a sense of induced, indirect flows. Accordingly, resource flows enable the various ways in which consumers desire or come to spend their time and should be analysed in respect to time-use' (Jalas 2005a: 132).

This passage echoes those economists who view time as the final form of utility (e.g. Zeckhauser, 1973). The final consumption of households is thus represented as a set of activities in which consumers make use of market delivered products as services, and activities are not, as in household economics, mere means for 'atemporal' ends. Consequently, the notion of efficiency also changes. Instead of being a ratio of functional output such as indoor temperature or vehicle kilometres per required inputs, efficiency is a ratio of lived or experienced time per the required inputs. To emphasise the paradigmatic change in view, I have previously argued that time in such a view is not 'spent' (as a resource), but 'made' (as the output of household production). In other words, everyday life consists of actions, in which we enact various embedded social roles, perform with our body, enjoy aesthetic experiences or good food, or simply just 'kill' time in pastimes. In all of these cases consumers do not optimise their time as a scarce resource in a Beckerian sense, but play out life in a condition of relative abundance. The relevant question is therefore: what are the required inputs of different ways of passing time?

The shift in focus can be illustrated by considering the IPAT formula (see Chertow, 2000), the derivatives of which are frequently used in societal analyses of energy use. The most simple formulation runs:

$$E = P * \frac{Y}{P} * \frac{E}{Y} \tag{1}$$

in which E = energy use, P = population, Y = economic output and E/Y = energy intensity of the economy.

The last 'technology' factor can be further decomposed into sectoral energy intensities in the following manner:

$$E = P * \frac{Y}{P} * \sum_i \frac{y_i}{Y} \frac{e_i}{y_i} \tag{2}$$

When using economic input-output tables (e.g. van Engelenburg et al., 1994; Biesiot and Noorman, 1999) that allow the allocation of all intermediary resource use to the items of final consumption, the aggregate energy use can be represented as:

$$E = P * \frac{C}{P} * \sum_i \frac{c_i}{Y} \frac{e_i}{c_i} \text{ (where C = Y)} \tag{3}$$

However, when taking a time use approach and viewing energy consumption from the point of view of everyday life, it is not the consumption expenditures that 'cause' energy consumption but rather various activities outside the hours of market work. The equation can be reformulated in the following way:

$$E = P * \frac{T}{P} * \sum_i \frac{t_i}{T} \frac{e_i}{t_i} \text{ (where T = total no. of hours)} \tag{4}$$

The last modification of the equation introduces the notion of the *energy intensity of activities* (e_i/t_i), which is the ratio of energy inputs per unit of time used in particular activities.[1] Hence, for example, the energy intensity of car-driving will be the ratio of direct and indirect energy inputs required by this activity per hour of driving. This formulation captures time-saving technological innovations particularly well. Appliances that reduce the time used for a particular activity (for example, cooking meals) may do so with an additional increase in energy consumption. Equally, the pace of life is also relevant. Driving a car more slowly, for example, will result in extended journey times and, in many cases, lowered fuel consumption.

The hours of work are one subject that the academics writing on sustainable consumption have focused on (Schor, 1991; Reisch, 2001; Sanches, 2005; Jalas, 2006). Reisch, for example, argues that the policy goal of sustainable consumption should include the pursuit of 'wealth-in-time' instead of increased material wealth. Sanne (2000) has connected working hours explicitly to the debate on rebound effects. He argues that only when efficiency improvements and productivity growth are coupled with a reduction in the average hours of work, can absolute levels of resource use be curbed, and the secondary, growth-generating effects of new innovations be absorbed. Equation 4 and the parameter T can be further modified to reflect this concern in the following way:

$$E = P * \frac{T}{P} * \sum_i \frac{t_i}{T} \frac{e_i}{t_i} \tag{5}$$

(where T = total no. of hours outside of market work)

In Equation 5, time is a form of ultimate utility, and the purpose of economic activity is to enable *non-market* time.[2] Whereas Equations 1, 2 and 3 use economic output (Y) to measure the output and the growth of this output, in Equation 5 growth is represented as increasing hours of non-market time. Technological progress is thus geared towards increasing non-market time and lowering the resource needs per unit of time of non-market activity.

To sum up, energy use in Equation 5 depends on:

- population;
- the amount of non-market time, which is the traditional growth effect;
- the composition of non-market activities, which is the structural effect;
- the needed energy inputs for specific non-market activities measured in, for example, megajoules per hour, which is the technology effect.

This everyday life perspective on energy use and the associated modification of the IPAT formula leads to some new questions. On the one hand, how is the capitalist drive towards increased amounts of goods and services fitted into everyday life? How is the problem of consuming the market output solved? How is the demand for time saving devices created and how is the scarcity of time maintained as an ideology? On the other hand, and bearing on the sufficiency debate, what are the activities that have a low energy intensity per unit of time and how do consumers establish and protect them from time saving innovations and instrumental rationalisation? How do proliferating, time-consuming commercial services such as golf fit in? Finally, and bearing specifically on the rebound debate, what are the potential impacts of energy-efficiency improvements for time allocation in everyday life?

These questions beg for a theory of time use. As obvious candidates, household economists have applied neoclassical economic theory to the time allocation of households and individual consumers. The work of Becker (1965) and Lancaster (1966) set out the perspective that economists should view the non-market activities of households as a form of production. That is, market goods are not the sources of utility *per se*, but rather factors of production in non-market activities. Most obviously, for example, kitchen utensils are combined with household labour to produce meals, but the Beckerian approach ultimately treats all consumption in terms of production. The optimal allocation of time between different activities will occur when the marginal utility of time use is same for each activity. Furthermore, the labour input depends on the wage rate and the available technological means to improve the efficiency of consumption.

The work of Staffan Linder (1970) is a relevant and direct application of this theory. He argues that the value of time will increase along with rising labour productivity in paid work and the consequent increases in wage rates. Hence there will be an incentive to increase the hours of work and to acquire market-provided goods and services to save time. He envisions a future 'harried leisure class' for whom time is extremely valuable and the members of which are hungry for technical remedies to reduce and fight persistent time famine.

Yet, I suggest a much more heterodox set of theorising on time use. While Linder's thesis seems to hold in that historically there has been a constant introduction of time saving devices and convenience consumables, as well as an increasing sense of harriedness, there is a danger of imposed rationality. The demand for time saving devices may not simply be a rational response to time famine. Time use surveys show for example that TV watching is the great winner in household time budgets. This observation is difficult to reason based on household economics; what are the technological innovations that have improved the productivity of TV watching? Equally illuminating, TV watching is not among the items that consumers list as their desired activities if only they had additional free time. Can it be that TV is just a handy tool for passing and consuming time? To say the least, it is evident that the 'decisions' of time allocation are made in conformance to the boundaries of what alternatives are available. Thus the satisfactory and customary schedule and structure of the day may be the guiding principle rather than an optimal, effective allocation of one's time.

Having established a time use approach to consumption, or an activity-based view of everyday life, there is a further point to make. Not all private consumption is linked to activities. Domestic heating or housing in general, for example, might be called infrastructure consumption that is not directly related any particular activities. Such infrastructure goods do not require the active and direct participation of consumers in order to be consumed (see the section on 'Data and methods' for further discussion). Hence, following the claim of Linder, aside from time saving devices, infrastructure consumption is feasible for the harried consumers. Additional square-metres of dwelling space, for example, call for no human participants and fit easily into the tight and stuffed time budgets of modern consumers. On the other hand, saturation of demand is much more likely in goods and services that require the active participation and attention of the consumers. These include, for example, personal transportation.[3] Such potential substitution effects and the market responses will be discussed in the following sections.

Time use rebound effects

This far we have established that private consumption consists of activities in which humans combine time with market inputs. Among these are the direct energy carriers, energy services such as district heating, and other goods and services with an embodied or 'invested' energy content. The latter include of course transportation services, but also manufactured goods and other commercial services, and the indirect energy demands of their production chains. Consequently, changes in the direct and indirect energy consumption of market production depend on changes in the allocation of time. This is most obvious in cases such as driving. The more hours of car travel, the greater the consumption of fuels, the greater the consumption of car-related services and the greater the range of industrial activities required to support these services. Apart from this principle of an activity-based view of consumption, we have made an important note that infrastructure consumption takes place without an apparent counterpart in the time budget.

In the following, I will demonstrate the use of these concepts while discussing so-called eco-efficient household services. The core idea of these services is that commercial providers of services enable a shift from owning various capital goods to purchasing the services they enable.

In such organisational arrangements, commercial fleet operators, such as car sharing companies, benefit from economies of scale and professional skills of operation, while consumers are able to acquire the level and type of services that best matches their current needs (see Heiskanen and Jalas, 2003 for a review of the debate). This debate has essentially been geared towards business actors and has called for innovative, more eco-efficient services or product-service-systems (von Weizsäcker et al., 1997; Mont, 2004). Yet, the environmental merits of such new service offerings remain unclear (Heiskanen and Jalas, 2003), partly because of the fact that outsourcing of domestic activities may generate a time use rebound effect.

Let us return to the lawn-mowing example. What type of effects might a commercial lawn-moving service induce? Following the introduction of such service, the time required for lawn mowing is removed from the time budget of the consumer/household. They now have additional time available and some new cost that represents the difference between the price of the service and the annualised capital expenses of the lawn mower.[4] In short term, the most likely response may be to cut expenditure on other activities. However, if similar services are used to substitute for other household activities, the outcome will be a surplus of time and a lack of money. In the long run it may be more likely that some of the lawn-mowing time will be allocated to market work in order to pay for the service. Furthermore, it is reasonable to assume that on average labour productivity is higher in specialised market work than in household chores and that thus not all of the available, additional time will be needed to acquire the income for the service. Available time can then be used, for example, to work more hours to increase the service level. People may, for example, buy more frequent mowing, engage with home entertainment, which has relatively low energy intensity (see Jalas, 2002, 2005a) or engage in high-consuming activities such as driving. The net result of the initial efficiency measure will, of course, greatly depend on these choices.

The secondary effects of a lawn-mowing service could be illustrated as in Figure 8.1. The axes represent energy requirements and time requirements. Hence, energy intensity of activities is the slope of the activity vectors. Optimistically, the service provider can provide the mowing service more (energy) efficiently, and thus the move towards using the service implies both a reduction of energy demand and the elimination of mowing time in the time budget. In order to pay for the service, the consumer may increase her working hours resulting in the vector L_S. Additional free time will be used to engage new non-market activities,

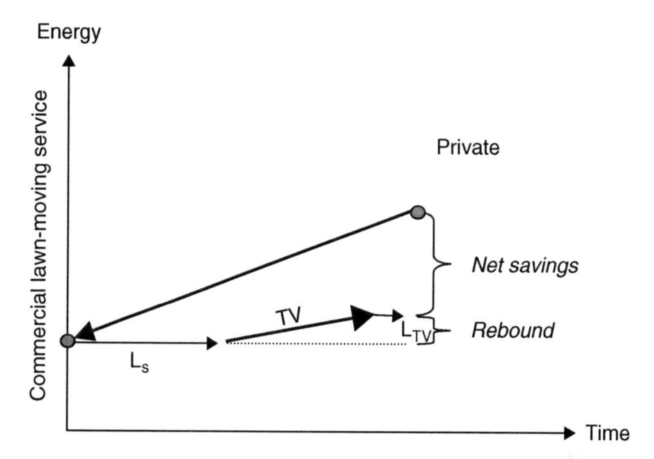

Figure 8.1 The time use rebound effects of introducing new household services: an example of a commercial lawn-mowing service with TV watching as the substitute activity

TV watching in this case (reflecting actual changes in time budgets) and possibly for additional labour activities – L_{TV} to compensate for the costs of this new activity.

The magnitude of the rebound effect would depend on

1. The impacts of the initial efficiency measure on consumption expenditure (the more costly the service the smaller the rebound).
2. The time savings created by the efficiency measure (the greater the time savings the larger the rebound effect).
3. The wage rate in paid work (the higher the wage, the greater the rebound effect).
4. The energy intensity of the secondary consumption activities and the costs of these activities (the more energy-intensive activities the greater the rebound effect).

The secondary effects leave much room for speculation. Perhaps most important is the rate that such services actually lead to additional hours of work. If and when there is an option to gradually increase the hours of work, it is not difficult to conceptualise instances of backfire, in which the secondary effects result in a net increase in energy consumption. This could be the case if the initial savings in energy consumption are small compared to the savings in time. Furthermore, if the time savings lead to new energy-intensive activities, a net increase in energy consumption is more likely. It is worth noting that driving is not the only way of using lots of energy in little time. Long-distance travel and, more surprisingly, hotel and restaurant services also score high on this ranking (Jalas, 2002, 2005a). Furthermore, with an eye towards increased obesity, it is reasonable to argue that eating in general is also an energy-intensive form of pastime in wealthy industrialised nations.

Previously, Binswanger (2001) has considered the rebound effects of time-saving technology using the theory of time allocation developed by Becker (1965). He identifies a double impact of such technology, arguing that, especially in transport, time saving (fastness) usually implies increased energy inputs per unit of delivered service (distance travelled). In addition to substitution between energy and time in the 'production' of the relevant service in transportation, time saving innovations will create additional demand for the same service. Furthermore, there is an 'income' effect as new activities can be fitted into the time budget.

At first sight, the rebound effect of outsourcing domestic activities appears to be positive (the secondary effect increases energy demand and decreases the net savings). Outsourcing of domestic activities such as lawn mowing or grocery shopping, transforms consumption from an activity-based category, which requires the participation and attention of the consumers, to a category which is market-based and is not constrained by limits or saturation in individual consumers' time budget. However, if coupled with more sustainable and time-intensive choices in everyday life, such as bicycle commuting or waste sorting, it is possible that the resulting secondary effects may reduce energy demand. At the same time, home entertainment and TV in particular may reduce the likelihood of significant time use rebound effects. TV seems to be a commodity that allows substantial time use while entailing little additional energy demand (Jalas, 2002).

The changes in time allocation, the emergence of new activities and the breakdown of traditional rhythms of everyday life are slow and complex processes. The puzzling case of increased TV watching is illustrative in this respect. The ageing of population contributes towards this change. Yet, at least according to Finnish time use survey data, an increase in TV watching has taken place among all age groups, despite being absent from the list of most preferred activities. Economic explanations of this phenomenon might include that TV watching is relatively easy, in that it requires no preparation, no skills, no effort and no social coordination. One might also note that the availability of this service has been greatly increased due to extended broadcasting times and more channels. Yet, all such speculation highlights the difficulty of establishing tendencies and forecasting future changes in the time budget. To that end, it may be more feasible to limit the rebound debate also in the field of time use to the more immediate impacts. One of the more easily accessible questions is whether energy-efficiency innovations imply the outsourcing of domestic activities and the transfer of activities from non-market actors to commercial actors. But an equally important question is to what extent individual outsourcing innovations lead to an increase in the hours of work and thus contribute towards a work-and-spend lifestyle (Schor, 1991) and a market bias in the provisioning of household services as a whole.

Market responses to time famine

One possible way to theorise the changes in time budgets is to consider time allocation from the point of view of market actors. If the prediction of a 'harried leisure class' made by Linder (1970) holds, consumers experience increasing time pressure and are inclined to substitute labour saving devices for domestic labour. Increased ease and convenience of consumption is a persistent target of new innovation in the consumer market. This applies not only to convenience food, but also to clothing (Velcro straps, wrinkle-free shirts), services (internet banking, quick lines in stores) and to many other categories of consumption (see Robinson and Godbey, 1997 for a record of various time saving products and practices). What is apparent is that new innovations have to compete extremely hard for consumers' time and attention. A time-saving innovation can fit into the course of everyday life much more easily that a time-taking innovation.

The scarcity of and competition over consumers' time changes the nature of innovations and product offerings. It may be that companies need to understand their product offerings and value propositions more and more in terms of time use and activity (Korkman, 2006). Where, on what occasions, by whom and in conjunction with what other activities is the product or service to be consumed? Historically, TV-dinners appear as a classic example of new, hybrid configuration of everyday activities. On the other hand, there are strategies or ways to overcome this time constraint or even to capitalise on it. Automated or programmable consumption is one cluster of convenience innovations. Washing machines, for example, perform work for us and consume water and detergents while not requiring much human labour. Equally, an automated lawnmowing service is an innovation that fits well into our busy everyday lives. The users of such goods do not need to be actively involved in their consumption; hence, automated consumption appears particularly favourable in the competition over consumers' increasingly scarce time.

Another category of consumption that the increased scarcity and value of time might foster is infrastructure consumption. These products or market offerings do not compete for the scarce time of consumers, but readily fit into the existing time budget. As already mentioned, additional living area is an example of energy-relevant consumption that is not directly related to activities. While this claim is not totally obvious since additional living area is often dedicated to, planned for or reasoned through new forms of activities such a home-gym or entertainment room, it remains the case that the commissioning of such space does not necessitate the reduction of some other activities. Equally, most heating and cooling services have no apparent counterpart in the time budget.

Finally, it is interesting to note that some time-intensive categories of consumption have been fostered even in the midst of apparent time pressure. A prime example of such consumption is the rapid increase in the sales of gardening products. Most of these products require considerable time and active human involvement. In other words, gardening, like many hobbies, is a time sink that is absorbing a lot of human effort. From an energy policy point of view, it is important to note that such time sinks may offer less energy-intensive ways of passing time and filling the normal days with satisfactory or highly rewarding activity. However, it is also apparent that such activities are the target market of a host of new service and product offerings and thus prone to become more goods-intensive and energy-intensive ways of spending time.

Data and methods

I have previously argued that a time use approach to consumption might offer new insights for energy policy. But, empirical tests of such insights are constrained by the available data. Data collection on time use has a relatively long history, but international comparative data is scarce. The history of time use surveys dates back to the early twentieth century (Jalas, 2005b). Since the 1960s there have been repeated representative national time use surveys in many industrialised countries. The harmonisation of data collection methods and data categories has benefited from the Eurostat guidelines for harmonised European time use surveys (HETUS) published in 2000. Aside from this effort, Jonathan Gershuny and colleagues have compiled a comparative set of international time use data from nineteen countries starting with the first national surveys in the 1960s.[5] The time use surveys typically consist of diary data and a rich set of background variables. However, what is problematic is that time use data and expenditure data are not typically collected from one single sample.

To measure the energy intensity of activities in terms of energy consumption per unit of time, data are needed on both time use and the direct and indirect energy consumption associated with each of the relevant activities. Thus, there is a need to link or merge the two separate data sets of time use and consumption expenditure. In addition, the notion requires that aggregates such as annual consumption expenditures and investments on durable goods are allocated to specific activities.

This process of matching consumption expenditures with the categories of time use has been a long-standing interest of household economics (Jalas, 2005b). Yet, this process is complicated; activity categories, to begin with, are a matter of dispute. The means–ends constellation is an empirical question; while at times cooking for example may be intrinsically meaningful and an end in itself, at other times it is simply a means for obtaining a meal. Should one, then, merge eating and cooking into the same category or maintain that these two issues are separate? Brodersen (1990) has elaborated on the matching process and suggested that resource use should be allocated to the immediately attached activities so that kitchen utensils, for example, are allocated to 'preparing meals' instead of 'eating', and renovation products to the very acts of performing the work. This implies a longer and more fine-grained list of activities. In addition, activities may link with more than one form of utility, as in the case of cooking, and 'multi-tasking' implies that various activities or aims overlap in the time budget. Furthermore, the question of infrastructure consumption is difficult; for example, to what extent should housing expenditures or energy demands be allocated to specific activities? Altogether, the process is arbitrary and requires empirical validation as opposed to pregiven categories and allocation principles.

However, comparisons of time use may nevertheless be revealing even though they necessitate problematic categories and time allocation rules. For example, in Finland it appears that the energy intensity of housework has risen during the 1990s (Jalas, 2005a). Finns, on average, use less time doing housework and spend more money and acquire more energy inputs to perform housework. This finding fits well with the prediction of the harried leisure class. Equally important, the aforementioned data operations enable rough estimation of the energy intensity of various non-work activities. This is one prerequisite for the assessment of the potential rebound effects of the alterations in the time allocation of households.

Conclusions

For many, the potential for energy-efficiency improvements appears promising and, even if the results await delivery, eco-efficiency thinking has set human ingenuity at the 'right' problem. Yet, for others the enthusiasm around eco-efficiency is hollow (e.g. Princen et al., 2002). Efficiency is not the right problem – instead there is need for a 'sufficiency' revolution. The notion of rebound effects has been at the core of this debate. It highlights the difficulty, or the fallacy, of maintaining two separate agendas, in which efficiency innovations are separate from the attempts to curb aggregate levels of energy consumption. If it is correct that improved energy efficiency leads to increases in demand, the agendas are not independent of each other. Instead, from a policy perspective, they should be merged, and any efficiency improvements should be reconsidered from the point of view of their wider impacts.

The rebound debate does reflect the concern for a dual agenda, and is characterised by a layered and shifting focus. The debate began with a single-service case. Later, authors such as Berkhout et al. (2000) proposed that the definition should span both the direct price-impact of higher demand for the same service as well as income effects which reflect the savings spent on other goods and services. Others have taken the debate to the other extreme and expanded the definition to cover the structural impacts of increased efficiency, including the contribution to economic growth (e.g. Sanne, 2000). In the narrow sense, the debate revolves around the price-elasticity of energy services, that is the degree that the generated savings will be used to acquire increased levels or amounts of energy-relevant products and services. In the broad sense, the debate surfaces as a political critique of capitalist economy and emphasises the tendency of markets to expand, and to acquire, transform and ultimately use up increasing amounts of natural resources.

Academics working with the latter perspective may view 'efficiency' as a flawed term, which masks the capitalist structure in which much of energy policy unfolds, and which creates false hopes of a technical remedy for issues that, for them, require political and institutional responses. In this chapter I have followed this line of reasoning. Hence, it appears that to serve the debate on sustainable consumption and to facilitate more radical forms of political critique, it is essential that the notion of rebound effects is allowed a broad definition. It appears necessary to include the transformative rebound effects that contribute to economic growth and market expansion regardless of the difficulty of empirical specification.

In this chapter, I have argued that a time use approach is one feasible way to capture transformative rebound effects. In previous literature, Linder (1970) and Binswanger (2001) have noted that the increasing value of time creates a demand for time-saving technology, which frequently is energy intensive. However, 'time-saving technology' is rather like the visible tip-of-the-iceberg, whereas many of the 'normal' energy-efficiency innovations also cause or contribute to changes in the time use of households. In this chapter I have taken the specific example of the outsourcing of domestic activities and the introduction of new eco-efficient household services and argued that this could lead to a net increase in energy demand. This speculative claim stems from a more solid proposal that the time use approach can unearth and illuminate a particular type of transformative rebound effect.

The empirical estimation of such transformative effects of energy-efficient innovations is difficult. As a first step, one needs to combine data sets on time use and energy use. The latter needs to include consumption expenditure data as well as information on the direct and indirect energy consumption associated with different household activities. However, provided that consumption and investment goods can be associated with particular time use activities, one should be able to translate the traditional rebound debate into a time-economy of non-market activities.

Notes

1. A similar methodology has been used by Juster and colleagues (1981) who studied the 'goods-intensity' of non-work activities. This measured the ratio of the value of market inputs and the time used to consume them. Gershuny (1987) has, for his part, focused on the labour-intensity of non-work activities, while measuring the amount of hours of paid work that are used (or created) in specific non-work activities.
2. The assumption that market activities yield no intrinsic process benefits can be challenged. In Equation 4 all categories of time use appear equal and hence there can be no growth effects towards 'better' time. The sharp distinction between non-market and market time in Equation 5 is thus merely a rough move to separate two polar qualities of time. Even if some contemporary professionals appear to get great satisfaction from their work, historically there is little doubt that the reduction of working hours has increased the quality of life of the majority.

3. The saturation of energy related demand is a claim that Lovins (1988) has used to argue that likely rebound effects are small. Yet, even in the case of personal transportation in which saturation in terms of time use is likely, this does not mean that the energy demands stabilise. The basic argument of Binswanger (2001) is that because of the limits of time or saturation due to time, transport technology develops towards increasing energy intensity be it measured in respect to time or travelled kilometres.
4. The assumption is that an outsourcing innovation such as a lawn-mowing service will increase the cost for households. Household services for maintenance and other housework seldom substitute very capital-intensive activities of households, but rather labour. In these cases, there will be no significant savings in capital costs. More radical innovations may save both money and time and thus create a more significant rebound effect, but a substitution between household labour and money is more common. Second-home ownership could serve as an example of the latter. In this case one might save both time and money by renting such a home instead of owning it.
5. See http://www.testh2.scb.se/tus/tus/ for available HETUS data and http://www.timeuse.org/mtus/ for the MTUS project.

References

Becker, G. (1965) 'A theory of the allocation of time', *Economic Journal*, 75: 493–517.

Berkhout, P. H. G., J. C. Muskens and J. W. Velthuijsen (2000) 'Defining the rebound effect', *Energy Policy*, 28(6–7): 425–32.

Biesiot, W. and J. K. Noorman (1999) 'Energy requirements of household consumption: a case study of the Netherlands', *Ecological Economics*, 28: 367–83.

Binswanger, M. (2001) 'Technological progress and sustainable development: what about the rebound effect?', *Ecological Economics*, 36: 119–32.

Brodersen, S. (1990) 'Reanalysis of consumer surveys: classification and method', in G. Viby Mogensen (ed.), *Time and Consumption*, Copenhagen: Statistics Denmark, pp. 273–90.

Chertow, M. R. (2000) 'The IPAT equation and its variants', *Journal of Industrial Ecology*, 4(4): 13–29.

Gershuny, J. (1987) 'Time use and the dynamics of the service sector', *Service Industry Journal*, 7(4): 56–72.

Goedkoop, M. J., C. J. G. van Halen, H. R. M. te Riele and P. J. M. Rommels (1999) *Product Service Systems, Ecological and Economical Basics*, The Hague: Ministry of Housing, Spatial Planning and the Environment.

Greening, L., D. Greene and C. Difiglio (2000) 'Energy efficiency and consumption – the rebound effect: a survey', *Energy Policy*, 28: 389–401.

Halme, M., C. Jasch and M. Scharp (2004) 'Sustainable home services? Toward household services that enhance ecological, social and economic sustainability', *Ecological Economics*, 51: 125–38.

Heiskanen E. and M. Jalas (2003) 'Can services lead to radical eco-efficiency improvements? A review of the debate and evidence', *Corporate Social Responsibility and Environmental Management*, 10: 186–98.

Hirschl, B., W. Konrad and G. Scholl (2003) 'New concepts in product use for sustainable consumption', *Journal of Cleaner Production*, 11: 873–81.

Jalas, M. (2002) 'A time-use perspective on the materials intensity of consumption', *Ecological Economics*, 41: 109–23.

Jalas, M. (2005a) 'The everyday life context of increasing energy demands. Time-use survey data in a decomposition analysis', *Journal of Industrial Ecology*, 9(1–2): 129–45.

Jalas, M. (2005b) 'Sustainability in everyday life – a matter of time', in L. Reisch and I. Røpke (eds), *The Ecological Economics of Consumption*, Cheltenham: Edward Elgar, pp. 151–71.

Jalas, M. (2006) 'Busy, wise and idle time: a study of the temporalities of consumption in the environmental debate', doctoral dissertation A-275, Helsinki: Helsinki School of Economics. Available at http://hsepubl.lib.hse.fi/FI/diss/?cmd=show&dissid=306

Juster, F. T., P. N. Courant and G. K. Dow (1981) 'The theory and measurement of well-being: a suggested framework for accounting and analysis', in F. T. Juster and K. C. Land (eds), *Social Accounting Systems: Essays on the State of the Art*, New York: Academic Press, pp. 23–94.

Korkman, O. (2006) 'Customer value formation in practice: a practice theoretical approach', doctoral dissertation, Helsinki: Swedish School of Economics.

Lancaster, K. J. (1966) 'A new approach to consumer theory', *Journal of Political Economy*, 74: 132–57.

Linder, S. B. (1970) *The Harried Leisure Class*, New York: Columbia University Press.

Lovins, A. B. (1988) 'Energy saving from more efficient appliances: another view', *Energy Journal*, 9: 155–62.

Meijkamp, R. (1998) 'Changing consumer behaviour through eco-efficient services: an empirical study of car sharing in the Netherlands', *Business Strategy and the Environment*, 7: 234–44.

Mont, O. (2002) 'Clarifying the concept of product-service system', *Journal of Cleaner Production*, 10(3): 237–54.

Mont, O. (2004) 'Reducing life-cycle environmental impacts through systems of joint use', *Greener Management International*, 45: 63–77.

Princen, T., M. Maniates and K. Conca (2002) *Confronting Consumption*, London: MIT Press.

Reisch, L. (2001) 'Time and wealth: the role of time and temporalities for sustainable patterns of consumption', *Time & Society*, 10(2/3): 367–85.

Robinson, J. and G. Godbey (1997) *Time for Life: the Surprising Ways Americans Use their Time*, University Park, PA: Pennsylvania State University Press.

Sanches, S. (2005) 'Sustainable consumption à la française? Conventional, innovative, and alternative approaches to sustainability and consumption in France', *Sustainability: Science, Practice, & Policy*, 1(1). Available at: http://ejournal.nbii.org

Sanne, C. (2000) 'Dealing with environmental savings in a dynamic economy: how to stop chasing your tail in the pursuit of sustainability', *Energy Policy*, 28: 487–97.

Schipper, L., S. Bartlett, D. Hawk and E. Vine (1989) 'Linking life-style and energy use: a matter of time', *Annual Review of Energy*, 14: 273–320.

Schor, J. (1991) *The Overworked American*, New York: Basic Books.

Schor, J. (2005) 'Sustainable consumption and worktime reduction', *Journal of Industrial Ecology*, 9(1/2): 37–50.

van Engelenburg, B. C. W., T. F. M. van Rossum, K. Blok and K. Vringer (1994) 'Calculating the energy requirements of household purchases: a practical step-by-step method', *Energy Policy*, 22(8): 648–56.
von Weizsäcker, E., A. Lovins and H. Lovins (1997) *Factor Four: Doubling Wealth, Halving Resource Use*, London: Earthscan.
Zeckhauser, R. (1973) 'Time as the ultimate source of utility', *Quarterly Journal of Economics*, 87(4): 668–75.

9
Rebound and Rational Public Policy-Making

Roger Levett

The very name 'rebound effect', and the metaphor it calls up, suggests something disconcerting and thwarting: you throw a ball, but it bounces back in your face, leaving you surprised, bruised – and back where you started in relation to what you were trying to achieve. Rebound is disconcerting because it undermines the common-sense assumption that if you pull a policy lever, things will move in the direction that you pull rather than the opposite – or, to put it a bit more formally, that the *overall* effects of a policy intervention will be consistent with its *immediate* effects.

This assumption is so basic to rational policy-making that we cling on to it despite the ineffectuality of action based on it to tackle 'wicked issues' such as climate change and transport. This chapter argues that an important reason for this is failure to come to terms with rebound, and the broader family of feedback effects of which it is a member. Progress requires a new 'systems literate' approach to public intervention. This is not difficult provided we drop some over-simplistic assumptions about the limits to public sector action.

The chapter starts by explaining why rebound is hard for public administration to assimilate, and why denying it is problematic. It then restates the classic Jevons rebound in systems terms, explaining that the occurrence and strength of rebounds depend on the context (technological, economic, commercial, social and policy) within which an efficiency improvement occurs, and drawing out implications for policy-making. This leads to an explanation in terms of political opportunism and of policy-makers' persistent failure to distinguish energy conservation from energy efficiency, illustrated by a snippet of the author's experience as a UK civil servant in the 1980s. The chapter then offers some principles for managing rebound and other feedback effects, including possible responses to ingrained vicious circles. The chapter uses UK examples, but the lessons are of general relevance.

Rocking the foundations of rational public policy

The expectation that there is some sensible, predictable, usable relationship between a policy intervention and its results is the indispensable basis for a rational approach to public policy and administration. It is a precondition for controlled experiments – where everything else is held constant so as to study the effect of one variable – to be possible or meaningful. It is enshrined in the economists' concept of ceteris paribus – 'if everything else stays the same', the conceptual key to their reductionist project of isolating causal connections to allow intelligent management of variables such as interest rates, borrowing, investment and tax to achieve desired ends.

It is also a prerequisite for Karl Popper's concept of 'piecemeal social engineering' (Popper, 2002) – the idea that an open society advances incrementally, pragmatically, empirically, by trying small interventions to tackle specific problems, seeing what works, and correcting and refining them. This was the intellectual basis of the Blair government's aggressively untheoretical approach, summed up in the slogans of 'evidence-based policy' and 'what matters is what works'.

Indeed, how could we manage without this common-sense assumption? If reducing the amount of energy machines and appliances take to do their work might increase the total amount of energy used instead of reducing it; if building roads might increase congestion instead of relieving it; if spending more on health services might leave the population sicker; if creating more jobs and expanding the economy might leave more people impoverished and deprived rather than fewer, how can we work out what to do to tackle climate change, congestion, health or poverty? Are we thrown back on the bombastic, mystical, untestable holism which Popper denounced as the root of totalitarianism?

Wicked issues

Unfortunately, there are 'wicked issues' including transport, health and poverty, which no amount of money spent on 'common-sense' interventions seems able to solve. 'What matters is what works' isn't working. Climate change has now joined them. Despite a long litany of actions and measures[1] the UK's greenhouse gas emissions are at best declining very slowly – and then only if, Enron-style, we put off-balance sheet the inconveniently rising emissions from the UK's international aviation and shipping, and industry in countries like China making goods for UK consumption. A frightening gap is opening up between the rapid reduction in greenhouse gases which the government recognises is necessary for climate security, and the minuscule, creeping changes that policy is achieving at best.

Understanding when, and why, the common-sense assumption breaks down, and how to respond, is an urgent need, and energy is the most important issue to do it in. This chapter argues that a proper understanding of rebound, and the family of feedback effects of which it is a member, is an essential part of the solution.

The rebound effect as an example of systems interaction

We start by understanding the rebound effect in terms of causal chains. Take the classic Jevons case of making a certain kind of engine more efficient in the sense of needing to burn less fuel to do a given amount of work useful to humans. The immediate consequence is that less fuel will be used to do the work those engines are already doing. But the second round consequence is likely to be more engines doing more work, because the increase in fuel efficiency has made each unit of work cheaper. If there is little extra demand for the work these engines do, energy use may drop; if there is a lot, the result could be many more engines using much more fuel.

Then there are third round effects. If the market for our engines grows, this is likely to stimulate competition to improve their fuel efficiency more. Shortages of fuel for the new engines, or people skilled in maintaining and operating them, could put a brake on their advance; however, in the longer term these could stimulate development of better fuel supply and training, which would further increase their attractiveness. And (at least for general purpose devices such as steam engines) the more familiar, reliable and cheap our engines and their supporting infrastructure become, the more attractive they will become which will tend to further increase their numbers.

And so on. A simple change has many potential consequences. Some of them may propagate, branch and loop around, either to thwart or neutralise the original change (negative feedback), or to reinforce and amplify it (positive feedback). Whether there is a rebound effect, and how strong it is, depends on the nature and strength of all the feedbacks that the original change triggers.

Significant features of rebound for policy

This account of rebound reveals several reasons why policy-makers find it indigestible.

Complexity and unpredictability

Whether we get 'just deserts' – efficiency improvements reducing energy use overall – or 'backfire' – efficiency improvements leading to no change or even an increase in energy use – depends on all sorts of factors in the broader context, ranging from the things the engine in question can be used for and the ways those things are currently done, to the availability of fuel.

This means that cases where a single intervention can satisfactorily manage some variable of interest to policy are the exception rather than the norm. Even the iconic example, the UK Bank of England Monetary Policy Committee's management of interest rates to deliver a politically set inflation target, only achieves its stark simplicity by leaving all the collateral consequences of rising or falling interest rates, from housing booms to manufacturing closures, to be palliated more or less effectively by other policy interventions. It is unfortunate that the British government is seriously proposing this as a model for reform of planning for major infrastructure projects (Communities and Local Government, 2007), an intrinsically complex and multifaceted policy challenge (Levett-Therivel, 2007).

Difficulty of 'evidence'

Another reason the rebound effect is indigestible to policy-makers is that the effects of a given change depend on lots of things that are apparently quite unconnected to that change, and which experts in that change may lack the ability, or the authority, to opine on. The doctrine of 'evidence-based policy' (Solesbury, 2001), ostensibly neutral and objective, actually skews policy appraisal towards overemphasising simple, direct, short-term consequences which are demonstrable, and sharply discounting longer-term indirect ones which are not.

Co-evolution of technologies and societies

Technologies and the social and economic systems they are embedded in co-evolve. Successive transport technologies have radically changed the size and form of cities and the lives lived in them. Railways and then lorries able to bring food in huge volumes quickly and reliably from

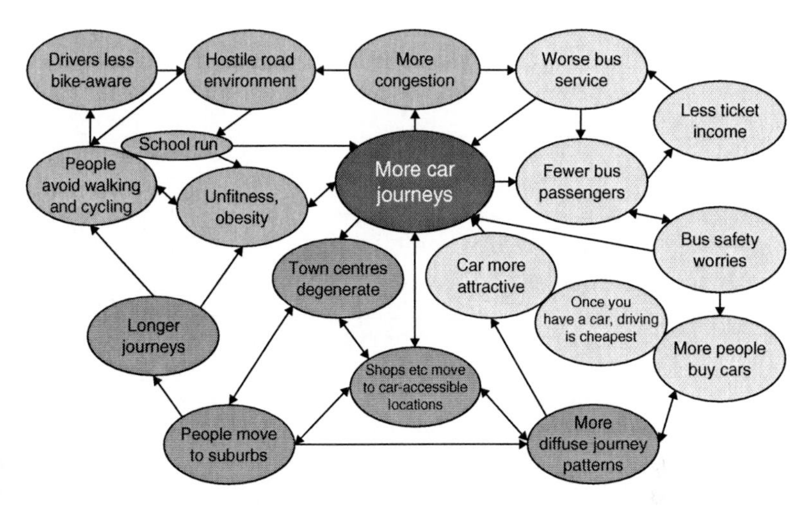

Figure 9.1 Vicious circles in transport policy

further away allowed cities to grow; commuter railways allowed people to work in office centres but live in leafy suburbs; cars now make it easier for them to travel outwards to office, retail and leisure parks. The spread of the car has helped mould the spatial structure of our cities so that it is now hard to live without driving. Figure 9.1 suggests how such co-evolution is continuing in transport.

A similar co-evolution has happened inside our buildings. Energy efficiency, rising incomes and falling real fuel prices leave most households and organisations able to heat their homes and workplaces to any desired temperature all year round without worrying about the cost. As a result many social groups expect to wear the same skimpy clothing indoors in all seasons, with warmer clothing only worn outdoors. Enclosed heated shopping and leisure malls, unknown in the UK until recently, have mushroomed to exploit the new dress habits, and further facilitate and entrench them. Now this has happened, brisk exhortations to turn down the heating and put a woolly on now run up against not only distaste for Puritanism, but also the difficulty many people would have mustering warm clothing suitable for indoor wear, and reluctance to stand out from social norms by looking baggy and frumpy. Starting from where we now are, if we want to evolve less use of energy to change the temperature of buildings, we will need to co-evolve habits and expectations of wearing different clothes in different seasons.

Irreversibility

The transport and warm buildings examples illustrate another irksome property of feedback: often it only works one way. We can drift into urban decay or heated buildings all too easily through the unintended interaction of lots of individual decisions and actions. But we can't get out again the same way. Households opting to buy cars in English towns in the 1950s were rewarded with effortless access to everything: it was only later that the cumulative effects of many others following them led to parking restrictions, congestion, one-way systems and so on. Fifty years later, anyone trying to opt out of the 'great car economy' faces a long wait at the bus stop, being splashed by all the car drivers who are delaying the bus, which even when it arrives can only offer access to a restricted and inferior subset of the town's amenities because the classier ones have scattered to places with good parking. The set of feedbacks shown in Figure 9.1 are like a lobster pot: easy to stray into, near impossible to get back out of.

Vicious or virtuous?

Whether a feedback effect is desirable or undesirable depends on the outcome one is concerned with. In the example, expanding the market for engines is good for energy efficiency: demand for more engines will increase the market incentive to make them more efficient. But the same feedback is bad for energy conservation: more engines means more fuel use. The failure to distinguish these objectives causes much confusion and ineffectiveness in energy policy, so we will consider it in more detail in the next section.

The philosopher Ernest Gellner (2003) pointed out that the word 'nobility' denotes both an admirable character trait and a hereditary upper class (he calls it 'bobility' to underline the general applicability of his point).[2] He argues that this curious conflation is neither a meaningless quirk nor a defect of language, but a device to enable 'nobles' in the hereditary sense to free-ride on respect and deference earned by people who are 'noble' in the other sense. What might appear just a muddle or inconsistency is 'functional' in the sense of serving a social or political purpose

The mischievous birth of the Energy Efficiency Office

The story of how energy efficiency gained policy prominence in the UK (which I witnessed first hand as a civil servant) shows Gellner's insight in action. The term was hardly used outside engineering circles before

1983 when Prime Minister Margaret Thatcher appointed Peter Walker as Secretary of State for Energy. Walker, a leading supporter of Edward Heath, her reviled predecessor, had survived in high office by confining his dissidence to oblique needling and teasing. On arrival at the Department of Energy he lit upon energy efficiency as a vehicle for this, rebranding the backwater 'Energy Conservation Division' as the 'Energy Efficiency Office', and changing 'conservation' to 'efficiency' in the names of all the schemes it ran.

Walker was no environmentalist. One of the small repertoire of riffs with which he delighted countless after-dinner audiences was on how, if gas, electricity or oil (depending on audience) had just been discovered, environmentalists would be up in arms calling for such a dangerous substance to be banned from homes and offices. His motivation for energy efficiency was purely political. Nobody could be against it. It offered unlimited photo-opportunities of the Minister as Father Christmas: opening energy-efficient factories, judging school poster competitions, congratulating the frail elderly occupants of the x thousandth house to be draught-proofed, presenting a van to a voluntary insulation group and such like.

Walker's enthusiasm for energy efficiency confused and wrong-footed environmental critiques of other policies. The plight of the 'fuel poor' (those unable to afford to keep warm because of low incomes and thermally poor housing) gave him a standing pretext for a jibe at the economic cruelty of Thatcherism whenever he wished one. A stream of case studies and advisory publications in which humble civil servants kept showing private businesses how much energy and money they were wasting provided running evidence that markets did not always function perfectly and that the public sector could improve things: heresy to Thatcher.

All this political fairy dust was made available to Walker by the switch from 'energy conservation', which he scorned for its miserable Puritanism, to 'energy efficiency', which was so beautifully aligned with prevailing discourses of managerial and business efficiency that even Thatcher loyalists who saw exactly what Walker was up to and hated it could never get, as it were, a clear shot at the kidnapper because he was holding the hostage too close.

Energy efficiency still politically 'functional' ...

Most subsequent UK energy ministers have milked the political benefits of energy efficiency (though few as astutely as Walker) and it has retained its presentational advantages, and therefore its ascendancy over 'conservation' in political discussion of the demand side of energy, ever since. In the late 1990s pragmatists including Michael Jacobs (1999) urged environmentalists to rally round the banners of 'ecological modernisation' and 'resource productivity' (which boils down to accelerating technical resource-efficiency improvements in the hope they can outpace consumption growth) because these words connected environment with two of New Labour's favourite aspirations, 'modernisation' and 'productivity', and with its project of shaking up the economy to deliver faster growth by exposing everything possible to the bracing winds of global competition. Jacobs urged environmentalists to distance themselves from 'sustainable development' with its unfashionable and politically disobliging ideological baggage of distrust of markets, growth and materialism – not because this was wrong, but because it would stop mainstream politicians listening.

The conflation of energy efficiency with energy conservation remains highly 'functional' (in Gellner's sense) to this day because it allows governments to appear to be acting vigorously on climate change while not doing anything that big business might regard as unfriendly or which might constrain economic growth and competitiveness. The emphasis of international climate negotiations on economic instruments, and especially on carbon trading, can be understood in this light. Whatever the theoretically predicted potential virtues of carbon trading schemes, schemes currently operating allow actual greenhouse gas emissions here and now to be excused by buying paper entitlements to the karma either of reductions that have already happened for other reasons (such as the closure of heavy industry in ex-Soviet states), or of notional future emissions avoided that might otherwise happen. For example in April 2007 the *Sunday Times* reported[3] that the Indian company SRF was to make about £300 million selling 'certified emissions reductions' (CERs) to European companies as a result of spending £1.4 million on a new plant to capture trifluoromethane, a potent greenhouse gas produced as a by-product of its operations, whose release is already illegal in Europe, This takes the decoupling of profitable commercial bustle from actual emissions cuts to new heights.

... but environmentally dysfunctional

Unfortunately the trick of conflating energy efficiency with conservation is highly dysfunctional for human security, because it prevents us recognising and confronting the inadequacy of energy efficiency as a means of avoiding the danger of climate change. Of course reducing energy use is not the same thing as reducing carbon emissions: this could be done by moving to low or zero carbon energy sources instead. But reducing the amount of energy they need to supply will certainly make it easier. Ignoring the possibility of rebound effects makes this conflation possible. Acknowledging rebound is thus urgently required as a reality check on evasive energy policy. Learning to manage rebound can help the design of effective carbon reduction policies.

Uncomfortable implications of feedback for policy

Before we can do so, we need to acknowledge some uncomfortable implications of feedback. Promoting energy efficiency is an 'obvious', common-sense way to reduce energy use. If roads are becoming more congested, the 'obvious' response is to add to capacity. If a bus or train service has lost so many passengers it is running at a loss, the 'obvious' response is to cut back to smaller, less frequent buses, stopping earlier in the evening. But these responses stoke up the feedback loops. Often the 'obvious' response simply makes the problem worse.

Choice can give us what we don't want

In the transport example above, every individual decision is perfectly sensible and rational for the individuals concerned, but they lead to results which nobody wanted: increased hassle and uncertainty for everyone in getting access to the things they want, but particular disadvantage for those without cars, and thus a deepening of inequality and exclusion; increased time, money, effort and space devoted to travel and transport. And, of course, increased carbon intensity: the historical evolution of road transport is, among other things, a gigantic rebound from better (including more fuel-efficient) cars (Commission for Integrated Transport, 2003; Levett, 2005).

Thus contrary to common sense and intuition, giving individuals 'more choice' does not necessarily lead to them getting more of what they want – or at least, not to all of them getting more of what they want. The reason is obvious once we think in systemic terms: each exercise of an individual choice alters the options available to others, leading them to make different choices, which in turn alter the options available to everyone else. In Figure 9.1, each individual choice to go by car instead of bus tends to make the bus less attractive to those still using it – a worse service, feeling less safe, and more delayed by car traffic. So, at the margin, each desertion of the bus makes more desertions likely, and we have a positive feedback underway. Thus it is possible to have a social rebound: more choice in movement leading to a situation where the choices people really want to make – to have cheap, easy, socially inclusive access to life's necessities and pleasures – are reduced (Levett et al., 2003).

Transport is not the only field where this happens. 'Parental choice' tends to mean articulate, confident families getting their (generally brighter and better behaved) children into what are perceived as 'good' schools. (Schools cannot instantly expand to meet demand, so 'parental choice' usually means popular *schools'* ability to choose the *pupils* they want.) These will tend to get better as a result, while the less promising and more problematic children of parents less able to work the system will get concentrated in the less desired schools, which will become worse as a result.

The problem cannot be prevented by setting geographical catchments, because then the better-off families can secure and entrench their advantage by outbidding others for houses in the catchment rather than by hiring private tutors. The tendency for feedbacks to entrench and exacerbate educational inequality can only be prevented in two ways: by *preventing* choice (for example by allocating children from all neighbourhoods evenly between all schools in town by lot), or by removing *motivation* for choice, by keeping all schools at near enough the same quality.

Government's job is to constrain choice

This means a government may be justified, or even mandated, to prevent or restrict individual choice in order to achieve public goods. This is shockingly subversive in a political culture where the extension of personal choice has become an almost universal aim. But it was well understood in the past. Social contract theory is based on the insight that some goods, such as law and order and the ability to do business with strangers of unknown morals, can only be achieved if each of us renounces some liberties – to assault, rob or cheat other people – in return for which the state stops anyone from exercising those liberties against us, by passing laws and maintaining effective police, courts and jails to enforce them (Mayo, 2007).

Government's power to restrict and coerce is justified if – and only so far as – the benefits the people get are worth the liberties they forgo. Social contract theorists formulate this in terms of a notional 'contract' negotiated between all the citizens, or between individuals and the state, hence the name. Of course the amount of state power people want changes between places and over time. Elections are, among other things, an institution for periodic renegotiation of the boundaries; party election manifestos are proposals for extensions or retrenchments.

Climate change and the social contract

Climate security is a 'public good': it can only be achieved if every country, sector and household 'does its bit'. It is an extremely high priority public good: virtually all others depend on it. But it is currently in grave jeopardy because in the absence of any effective collective constraint, individuals and companies continue to make decisions about energy use which are perfectly sensible for their private interests, but lock us in to collectively catastrophic climate change. Species survival demands a prompt renegotiation of the social contract to add carbon to the issues on which government is empowered and expected to interfere with individual license (Mayo, 2007). A 'systems-literate' approach to policy is a prerequisite for this. The next section outlines some elements of this approach.

Principles of systems-literate policy

Positive feedback provides a free ride; negative feedback or rebound a counter-current to overcome. Systems-literate policy-making is the art of making feedbacks work for you, not against you: to approach the social-economic-environmental system like a sailor using winds and tides, or a glider pilot seeking out thermals, rather than like a motor boat or plane using brute force to fight against currents. This section sketches some principles to help achieve this.

Avoid encouraging unwanted feedbacks

The first is simply to *avoid actions and responses that help unwanted feedbacks develop*. An example would be to refrain from providing additional parking or road capacity, and just stand back and allow the negative feedbacks of rising frustration with gridlock and inability to park set physical limits to the growth of traffic in a town. Increasing capacity is so self-evidently the 'sensible' response to increased congestion that this has until now been exceedingly difficult to propose as a deliberate response – even when the congestion has just reappeared, at a larger scale, on a road that was only built or widened a few years ago to deal with a previous congestion problem, or where road building is itself damaging, expensive and unpopular. Moreover it can easily create undesirable knock-on effects of its own, for example loss of visitors and trade to competing towns or out-of-town locations. However, it is striking that towns with historic centres that have resisted traffic 'improvements' by being physically incapable of accommodating wide roads and big car parks have tended to thrive (for example in 2006 the top four of claimed 'best places to live in the UK', by the UK TV Channel 4, were Winchester, Horsham, Tunbridge Wells and Harrogate).[4]

Raise prices to offset efficiency improvements

A second principle is *forestall or neutralise unwanted feedbacks at source*. For example, the 'classic' Jevons rebound of increased energy efficiency leading to increased energy use could be neutralised by raising the cost of energy at the same rate that efficiency improved, so the apparent cost to consumers of each unit of energy services stayed the same. Of course it cannot be quite that simple. If tax rose in line with the increased efficiency of new houses (or cars, or whatever), owners of older ones would have to pay more as a result of technical advances they got no benefit from. But if the higher tax were (somehow) levied only on the owners of the new, more efficient, products, this would remove manufacturers' incentive to improve efficiency, because the higher tax would claw back the benefit customers would otherwise get from buying them.

In any case if the aim is to prevent increased energy use, our new anti-rebound energy tax would have to rise in line with incomes or spending power, to prevent the perceived affordability of energy services from rising. But this would increase the disadvantage of people whose incomes or spending power was rising at less than the average rate. Moreover there are lots of different energy using products, whose energy efficiency is improving at different rates (including zero). It would not be practicable to levy different tax rates for energy used in (say) washing machines versus room heating versus DVD players. Any practicable scheme for using energy taxes to neutralise rebounds from energy-efficiency improvements would have complex redistributive effects between different people and energy using products, and have more effect on some needed to neutralise rebounds, and less on others.

The ideal of engineering taxes to precisely neutralise energy efficiency rebound effects is therefore unrealistic. However, the broader point remains valid that high and rising energy prices will tend to strengthen those feedbacks that will tend to reduce energy use, and weaken those that tend to increase it. For example, the decay of urban areas presented in the transport diagram has not gone nearly as far in the UK as it has in North America and Australia, partly because transport fuel has always been much more expensive (there are many other reasons too, as will be discussed later). This could be expressed as an alternative second general principle: *use taxes (and other economic instruments) to tilt the playing field in favour of desirable feedbacks and against undesirable ones*. (The 'level playing field' frequently invoked by special interest lobbyists, often to resist regulation, is a myth. There is no 'objectively' correct or 'fair' relationship between, for example, the cost of energy and other factors of production. It is always and intrinsically a matter for policy choice.)

Cost structures

One contribution to the vicious circle in transport is that much of the cost of owning a car is 'sunk': you have to buy it, tax it, insure it and take an annual hit on depreciation before you drive it a mile, but then each mile you drive costs only the fuel and an apportionment of tyre wear and servicing. As a result, once a household has to acquire a car, they have an incentive to use it for as much of their travel as possible, because at the point they decide between car and bus, the car will almost always appear cheaper. A different cost structure, with costs shifted off owning cars to driving miles (for example, car taxes replaced by higher fuel taxes) could reduce this perverse incentive. So could road-user charges, parking taxes, or car insurance charged per mile driven rather than per year.[5] Higher charges for residents' parking for gas guzzlers, while an indirect way to reflect environmental damage, might yet be effective as ways to project concerns or jolt people into different behaviour. Car sharing schemes achieve the same effect by structuring their charges as a relatively small annual fee and relatively high charges per mile driven (though the extra element of an hourly charge, needed to give members an incentive not to keep them longer than needed, slightly confuses the signal).[6]

The problem could also be tackled by reforming the cost structure of public transport in the opposite direction: making public transport a more attractive alternative for car owners by reducing its cost at the point of use (i.e. the price of tickets) and making up the shortfall through fixed charges such as local taxes. Unfortunately the long shadow of the emotive Thatcherite slogans of 'eliminating subsidy' and 'forcing loss-making public services to balance their books' still block the political space for rational and pragmatic discussion of the best combination of fixed and trip-related charging for public transport to achieve public policy goals including removing perverse incentives to drive.

The same principle could be used in different contexts. For example, if domestic energy suppliers were required to sell every householder enough power to keep their home warm at a low, loss-making cost per unit, and to charge a higher cost per unit above the threshold, provided the prices were set carefully the result would motivate energy suppliers to promote energy *efficiency*, and motivate consumers to use it to achieve energy *conservation*. Energy suppliers would have a commercial incentive to subsidise energy-efficiency measures in houses, to push down the amount of power they would have to sell at a loss. Householders facing a steep cost for their marginal consumption would bother to minimise the amount of power they used at the higher tariff. While there may be practical difficulties in such proposals, they all illustrate a third principle, namely *structure incentives to make the desired results easy and attractive at the point people make decisions.*

Structure markets to align commercial benefit with public goods

Leasing rather than selling products can replace the traditional perverse incentive for manufacturers to build in obsolescence with a benign one to maximise product life and minimise resource costs. Energy service contracts (Sorrell, 2006) where an energy company contracts to give a customer an agreed level of energy services – for example comfortable, well-lit offices – and makes its profits by applying technologies to reduce the amount of energy needed to deliver those services, is a further example of a fourth principle: *design commercial and regulatory structures and institutions to align commercial benefit with desired environmental outcomes, and to eliminate perverse feedbacks.*

In the 1970s and 1980s when American energy utilities were geographical monopolies, public sector regulatory commissions were established to protect customer interests. One way they did this was to only allow the utilities to recoup the costs of investments which the regulators were satisfied were the most cost-effective way to meet energy demands. No new nuclear power plants were built for decades because they could not pass this test. Some regulatory commissions, notably California's, took the principle a creative step further, by only authorising investments in *increasing energy supply* if the utility could prove these were cheaper than measures to *reduce energy demand*. Often demand reduction was cheaper. Then the regulatory commission authorised the utility to offer customers heavily subsidised or even free energy conservation measures, and recoup the costs from fuel bills, because this was cheaper than building the power stations that the conservation measures saved the need for (Association for the Conservation of Energy, 1984).

Behaviour change as the driver

The point of all these measures and principles is to change behaviour. People are only economically rational to a limited extent, so 'getting the prices right' by (for example) reflecting economic valuations of externalities is less important than understanding when people make the decisions that affect their energy behaviour, and what will make them adopt more sustainable options. The annual increase in motor fuel taxation was beginning to work at the point the government abandoned it, because it created an expectation that fuel prices were going to continue to rise, which was starting to make people think about fuel-cost implications when considering buying a car, or moving house or job. These longer-term decisions have much greater potential to save energy than day-to-day travel or energy behaviours, in which most people tend to follow routines for years at a time until jolted out of them. Most people do not recalculate their 'utility functions' in the light of the latest energy prices.

Thus a fifth principle is, *in designing interventions, start from the behaviour changes desired*. This requires attention to 'trigger points'. The latest government guidance on company travel plans offers a sensible example: 'Staff reliant on [bus] services that stop after peak hours may fear getting stranded at work if they have to stay late unexpectedly. The solution is to offer a guaranteed ride home, which is now a commonly used safety net for car sharing schemes.'[7]

As another example, anxieties about being 'ripped off' by contractors can create a big disincentive to energy conservation investments, but accreditation of those contractors by trusted local authorities or community organisations can offer an effective response.

Preserving virtuous feedbacks

Continental European cities with a historic legacy of high density, attractive centres with excellent amenities, urban living, good walking and cycling access and public transport have self-reinforcing feedback loops which are the opposite of those shown in Figure 9.1. Such places can prevent the vicious circles of the diagram from getting established, or have been able to contain and thwart them with strict spatial planning policies and car restraint. A sixth principle of systems-literate intervention is therefore to *recognise where virtuous circles are working, avoid unintentionally disrupting them, and give support to keep them functional*. This is like the Boy Scout shown in 1950s physics textbooks, able to hold a telegraph pole balanced upright with one hand, by correcting any drift out of the vertical before it gains significant momentum.

Getting from vicious to virtuous

However, once the vicious circle has taken hold, gentle measures cannot reverse it. In most of the UK, subsidised buses run nearly empty; developers vote with their feet when planners try to make them accept places with poor roads or restricted parking; families locked into a cat's cradle of time-critical criss-cross journeys to get everyone to and from schools, workplaces, shops and recreations all in different places react with derision to suggestions they should use sparse and inflexible buses, and with fury to any attempt to curb their driving. Rational spatial patterns, good public transport and car restraint could all support each other. But none of them works without the others already working. This section discusses how the problem of interlocking vicious circles can be tackled. There are three possibilities.

Start from scratch

Conceptually the simplest way to dodge a vicious circle is to build a new system with the desired feedbacks built in from the start. Vauban, a new 'sustainable urban quarter' built on the outskirts of Freiburg in southern Germany between 1998 and 2005 is a good example.[8] Extremely high mandatory insulation standards, a requirement to buy heat and power from a wood-chip plant on site, good social infrastructure on site, a tram line down the middle of the neighbourhood and cars only allowed on the edge (and only on payment of a swingeing extra charge for a parking space) are among the features that ensured that residents adopted mutually reinforcing sustainable low energy life patterns from the start. On a massive scale, a complete new city, Dongtan, is being built near Shanghai, aiming for carbon neutrality.[9]

Such a bold scale of action would seem beyond the UK. However the government is committed to build several Dongtans-worth of new housing across England over the next twenty years or so. Most of this will be squeezed incrementally into existing urban areas, and extensions to them, with much greater cost and difficulty over infrastructure, environmental resources and effect on neighbours than the Chinese face with former agricultural land at Dongtan, and, currently, no intention to set the kind of frameworks and conditions which are intended to make Dongtan virtually carbon neutral. The difference between China and the UK is thus not the scale of development, but the will and readiness to intervene to exploit large-scale development to make a step change toward sustainability.

Flip the system

A second possible response to vicious circles is a big and strong enough intervention to flip vicious to virtuous. Ken Livingstone as Mayor of London achieved this with the congestion charge (Transport for London, 2007). Supported by massive investment in buses, this has created a new virtuous circle in which less private traffic in the charge zone has enabled buses to move faster, providing a better service, enabling more people to stop driving, and allowing reallocation of road space to public transport, cyclists and pedestrians, and for public-realm improvements. However, this required both supportive circumstances and an unusual level of political imagination and courage.

Create a virtuous island in a vicious system

A third response is to create a small bubble or island of virtuous feedback in an otherwise vicious context, and then help it grow and propagate until eventually the whole system has flipped. The eco-tipping points project[10] has researched a number of examples. In a New York example, a group of citizens clearing the rubbish off a derelict site and making it into a community garden transformed a magnet and amplifier of alienation, anxiety and crime into the opposite: a generator of community cohesion, optimism and engagement, which discouraged crime by bringing more law-abiding people out on to the streets. As in other eco-tipping point case studies, once the first isolated example had been achieved by a few people motivated enough to get physically stuck in to do something new, others saw that it worked, and copied it, and over a period hundreds of gardens were created, and the whole tone of neighbourhoods transformed.

Conclusion

This chapter has suggested that it is fertile and productive to think of rebound effects as members of a family of unintended consequences whose defining characteristic is their dependence on feedback effects. If we can acknowledge and manage feedbacks more effectively, we may be on the way to deal with a range of currently intractable problems. This requires a major upgrading of our approaches to public policy and management.

Important policy problems involve a range of causal pathways, some of which may be counterintuitive. Technological and social systems intertwine in complex ways, making precise prediction of the results of any intervention impossible beyond the very short term. The intrinsic indeterminacy of complex social systems means the connections are not provable. The attempt to justify every decision with 'evidence' may privilege the trivial over the important and miss exactly what matters most. Beliefs, values and habits are part of the mix: making a behaviour change economically advantageous is neither necessary nor sufficient to make people do it.

All this underlines Popper's calls for a tentative, experimental approach to public interventions. But it means that the kind of small, discrete, incremental experiments he advocated – change a single variable such as a tax rate and see what happens – are likely to miss the point. We need instead to experiment boldly and decisively, such as with the design of whole neighbourhoods, towns and cities. We need systemic rather than piecemeal social engineering, both enabling sustainable behaviours and restricting unsustainable ones as two sides of the same coin. For example, energy-efficient buildings *and* taxes or tariff structures that sharply penalise unnecessary energy use; guaranteed excellence in local schools, clinics and other social services *but* no entitlement to use any other than the nearest one except in very special circumstances; excellent public transport and walking/cycling facilities *and* high fuel, road use and parking charges for private cars. And soon: climate change will not wait.

Notes

1. For example see the *UK Climate Change Programme 2006*: http://www.defra.gov.uk/ENVIRONMENT/climatechange/uk/ukccp/index.htm
2. 'Bobility is a conceptual device by which the privileged class of the society in question acquires some of the prestige of certain virtues respected in that society, without the inconvenience of needing to practise it, thanks to the fact that the same word is applied either to practitioners of those virtues or to occupiers of favoured positions. It is, at the same time, a manner of reinforcing the appeal of these virtues, by associating them, through the use of the same appellation, with prestige and power. But all this needs to be said, and to say it is to bring out the internal logical incoherence of the concept – an incoherence which, indeed, is socially functional' (Gellner, 2003: 39).
3. http://www.timesonline.co.uk/tol/news/uk/article1687531.ece: 'Indians make cool £300m in carbon farce'.
4. http://www.channel4.com/4homes/ontv/best&worst/2006/best.html

5. Although a pioneering scheme by major insurer Norwich Union offers the cheapest rates for motorway driving. This is commercially sensible because this is less accident prone, but could be environmentally counterproductive
6. See http://www.citycarclub.co.uk/
7. DfT, *The Essential Guide to Travel Planning, 2007*: http://www.dft.gov.uk/pgr/sustainable/travelplans/work/makingptpworkessential
8. See www.vauban.de
9. See http://postcarboncities.net/taxonomy/term/35/9
10. http://www.ecotippingpoints.org

References

Association for the Conservation of Energy (1984) *Lessons from America*, London, ACE.

Commission for Integrated Transport (2003) *10 Year Transport Plan: Second Assessment Report*, London: Commission for Integrated Transport.

Communities and Local Government (2007) *Planning for a Sustainable Future*. White Paper, London: TSO.

Gellner, E. (2003) *Cause and Meaning in the Social Sciences*, London: Routledge.

Jacobs, M. (1999) *Environmental Modernisation*, London: Fabian Society.

Levett, R. (2005) *Planning for Urban Sustainability: an Ecosystems Approach*, London: RTPI.

Levett R., I. Christie, M. Jacobs and R. Therivel (2003) *A Better Choice of Choice*, London: Fabian Society.

Levett-Therivel (2007) *Deconstructing Barker*, London: CPRE.

Mayo, E. (2007) *The Environmental Contract: How to Harness Public Action on Climate Change*, London: National Consumer Council.

Popper, K. R. (2002) *The Open Society and its Enemies*, Volume 2: *Hegel and Marx*, London: Routledge.

Solesbury, W. (2001) *Evidence Based Policy: Whence It Came and Where It's Going*, ESRC UK Centre for Evidence Based Policy and Practice: Working Paper 1. See: http://evidencenetwork.org/cgi-win/enet.exe/pubs?QMW

Sorrell, S. (2006) 'The economics of energy service contracts', *Energy Policy*, 35(1): 507–21.

Transport for London (2007) *Central London Congestion Charging Impacts Monitoring Fifth Annual Report*. See: http://www.tfl.gov.uk/assets/downloads/fifth-annual-impacts-monitoring- report-2007-07-07.pdf

10
Avoiding Rebound through a Steady-State Economy

Jørgen S. Nørgård

The debate on the *rebound effect* as presented in most chapters in this book is based upon experience from the *past* more than visions of the *future*. The analyses are dominated by conventional economic theory, which implicitly assumes insatiable demand for energy services. Material consumption is considered to be limited primarily by productive capacity with little concern for ecological costs and limits. In such a development aiming at unlimited growth it would from a long-term environmental perspective be close to irrelevant to reach for more efficient use of energy, since it would only buy some time. From this perspective, the environmental problem with the rebound effect is not the higher energy efficiency, which pushes towards *lower flows* of resources through the economy, but rather the conventional economy which rebounds the savings, because of its quest for *higher flows*.

In this chapter, I shall take the rebound debate further by discussing the possible role of energy efficiency in a sustainable economy that is based on the notion of 'sufficiency'. The assumption is that globally we need to achieve a 'steady-state economy'. Considering the urgent need for better material conditions in many parts of the world, the transition towards a steady-state economy needs to begin first in the affluent countries, including the Nordic countries from where most of the information in this chapter is drawn. The politicians in these countries are not seeking a steady-state economy, but some social and cultural traditions may provide prerequisites for such a society, including public attitudes towards low birth rates, equity, consumption and work. Although this chapter presents a Nordic perspective, the options and trends described are relevant worldwide.

It is assumed that absolute reductions in energy consumption are desirable, since most energy supply options involve environmental problems.

While renewable energy sources are generally more environmentally benign than fossil fuels and nuclear, they nevertheless constitute a very direct intrusion on nature, as is already apparent where hydropower and biomass are used intensively.

Efficiency in growth economies

The economic growth experienced in parts of the world for the last two centuries has been driven in part by 'rebound effects' arising from substantial increases in efficiency (or productivity) in the use of various resources. For example, only a small fraction of the increase in labour productivity has been converted into less labour input; the rest has rebound as more production. Rebound effects from the more efficient use of energy have also spurred economic activities. However, it would be wrong to ascribe all the growth in the use of energy service to energy efficiency alone. While energy use for lighting, for instance, has increased *alongside* improvements in energy efficiency, it is not only *because* of those efficiency improvements that energy use has increased (Herring, 2006a). The proportion of the growth that is due specifically to improved energy efficiency is what the rebound debate is largely about. But as will be discussed in this chapter, in a future steady-state economy the outcome of such improvements may be quite different.

The delusion of decoupling

In the 1980s it became popular to claim that energy consumption was – or in the future could be – decoupled from economic activity, as measured by GDP. However, the decoupling of energy consumption from economic activity is largely a statistical delusion. Physically the economy has very real links to energy consumption, although not necessarily to carbon dioxide emissions. Every economic activity requires some energy consumption and all energy consumption is rooted in some economic activity, be it on the consumer or producer side. The observation that the two parameters can grow at different rates, which over history is quite normal, does not imply any decoupling. Whether a car is running in first or second gear, there is still a coupling between the engine and the wheels, and speeding up the engine will speed up the car. Unfortunately, the notion of a decoupling has served as peacemaker between environmentalists and growth-oriented politicians by conveniently exempting economic growth of any responsibility for environmental problems.

An example often used in support of decoupling is the fact that, over the last thirty years, energy consumption in Denmark has remained approximately constant while GDP has grown by more than 70 per cent. This is partly the result of government policy to promote the more efficient use of energy, beginning with the oil crisis in 1973 and including energy taxes, subsidies for energy saving investments, building regulations and information campaigns. Energy savings have been achieved by both end-users and suppliers, with the latter replacing electrical space heating and individual boilers with district heat from combined heat and power generation. The reduction in energy consumption for space heating has, however, been offset by the rapid growth in car traffic where efficiency gains have been modest. In total, economic development has eaten up what could have been a reduction of more than a quarter of energy consumption over the last thirty years, if there had been no growth *or if there had been no coupling to the growth.*

In addition, the above figures follow the Kyoto Protocol in excluding energy consumption used for air travel and merchant shipping. On a per capita basis, the latter is exceptionally high for Denmark and if included, Danish per capita energy consumption and carbon dioxide emissions are found to have *increased* by more than 50 per cent over the last thirty years – to become one of the largest in the world.

Given continued economic growth and the associated assumption that people can never have enough goods and services, energy savings from better technology can only temporarily offset the drive towards higher energy consumption (Nørgård, 1974). Hence, in an ever growing economy, relying on technology alone to provide a lasting and significant reduction in energy consumption will trap us in an eternal Sisyphean task.

A political predicament

The contradictions between pursuing sustainable development and yet insisting on continued economic growth is illustrated by a comment from a parliamentary spokesperson from one of the Danish governing parties. When asked whether the government's policy would achieve energy savings, he replied: 'We can reduce energy consumption, we can reduce our transport, we can buy fewer goods with large energy content. But all this also implies a reduction in the level of activities, and we are not interested in putting the brakes on the general economy, as a consequence of people's concerns' (Nielsen, 2006). This is a surprisingly frank expression of the dominant political preference. When the development path requires a choice between growth and sustainability, the former is invariably preferred.

The policy-makers' predicament partly explains the half-heartedness of many energy saving policies. For example, information campaigns never encourage savings of the *indirect* energy consumption as embedded in all consumer goods and services. For instance, policy-makers could urge consumers to save energy by postponing replacement of furniture, clothes or other durable goods, which would be an obvious step towards environmental sustainability (Nørgård, 2006a). However, this would conflict with the promotion of increased consumption.

Private businesses aim to make profits and consequently have an interest in using resources efficiently in their business of providing products to consumers. On the other hand, they also benefit from higher sales if customers use the products wastefully. With this in mind, it is striking that governments worldwide frequently turn energy saving responsibility over to energy suppliers! Although there are examples of energy supply businesses taking this responsibility seriously, this model introduces an organisational conflict of interest – like letting the wolf watch the sheep. Not surprisingly the suppliers' strategies are often half-hearted and actual savings have been hard to detect. A particularly unfortunate example is an information campaign by one Danish power utility, which recommended that customers switch to the three times more efficient low energy lamps and then suggested using three times as much light. In other words, they were encouraging a 100 per cent direct rebound effect (Nørgård, 2000).

A better political option for reducing energy consumption would be an independent agency with the sole obligation to save energy. This option was chosen by the Danish government in 1996, after they lost patience with power utilities that had refused to subsidise the conversion of electrically heated buildings to district heating. The government then established the Danish Electricity Savings Trust (Nørgård et al., 2007).

The conflict between continuing economic growth and a shift towards a sustainable development is well illustrated by Brookes (1990) who advocates the view that improvements in energy efficiency will in general lead to an increase in energy consumption. Brookes does admit that energy efficiency could reduce energy consumption in some circumstances, but warns that this would be at the expense of a loss in economic output (Brookes, 1990). The latter point is correct, but should not be taken as a warning, but rather as a hope, since maintaining human well-being with a lower economic output can and should be considered as a realistic goal for the future. Like many other conventional economists, Brookes confuses a *large* output with a *good* economy. This perception is a key problem in debating sustainability, not to mention achieving it.

Misplaced end-use efficiency?

Does this mean that our efforts to develop energy-efficient technologies are in vain from an environmental point of view? Certainly not, but putting faith in technological efficiency improvements as *the first and sole means* to reduce energy consumption would be very short-sighted. Nevertheless, the extent to which the potential for technological efficiency improvements are utilised will determine the level of energy consumption, environmental impact and human welfare we can ultimately choose from. Experience has shown that, first, the energy savings from efficiency gains are being eroded by various rebound effects, and second the efficiency gains themselves are being used as a pretext for doing nothing to change the basic driving forces behind the problem – namely the growth in population and consumption. A more valid approach would be to give first priority to beginning the long-term economic, social and psychological transition towards a steady state, as described below, and then along the way gradually implement more energy-efficient technologies (Nørgård, 1974).

Sustainability targets and definitions

The contemporary problems of climate change and depletion of oil reserves ('peak oil') are examples of trends that were largely anticipated by the path-breaking 1972 report, *The Limits to Growth* (Meadows et al., 1972, 2004). This study used a dynamic model to analyse the interconnection between important global parameters and the consequences of different development options, some sustainable and some unsustainable. The continued pursuit of economic growth was in the model found to be unsustainable and likely to result in severe problems in the twenty-first century, for instance with pollution and resource shortages. After being ignored and criticised, often on erroneous grounds, the basic messages of *The Limits to Growth* report now seem more relevant than ever (see for instance Simmons, 2000). Unfortunately, so far we have missed thirty-five years for a gradual transition towards a sustainable economy.

While the 1987 Brundtland report introduced the concept of sustainable development, the proposed definition is open to varying interpretations (WCED, 1987: 43). An alternative approach is offered by Herman Daly, who borrows the economic definition of income, namely the maximum amount that a person or a community can consume over some time and still be as well off at the end of the period as at the beginning

(Hicks, 1948). From this, it can be deduced that an environmentally sustainable economic development is a development which does not reduce the development options of the future by deteriorating the natural environment (Daly, 1990, 1996). This implies that the exploitation of renewable resources and emission of pollutants must be kept below nature's capacity to provide and neutralise them respectively. Non-renewable resources should in principle not be used, but to the extent that we accept substitutability between man-made and natural capital, they can be used but only at a rate below that of establishing appropriate substitutes such as energy conservation and sustainable renewable energy supplies. In any case, these rules should be seen only as *necessary* biophysical conditions for the sustainable use of the natural environment, not as *sufficient* conditions.

Sustainability and the steady state

In the infancy of economics as a science, the British philosopher and liberal economist John Stuart Mill in 1848 expressed his concern for the environmental consequences of economic growth. He hoped that people 'would be content to be stationary, long before necessity compels them to it' (Mill, 1900: 264). Since that time, several economists have argued that growth should be considered a transition to maturity where a stationary or *steady state* is reached. Two Nobel laureates in economics have separately expressed the need to stop growth: Trygve Haavelmo by saying that 'further growth in the rich countries would be a terrible thought. It is inconsistent with the environment' (Vermes, 1990), and Jan Tinbergen stating that in order to reach sustainability the highest priority is to 'permit no further production growth in rich countries' (Tinbergen and Hueting, 1991: 41).

According to Herman Daly (1977: 17, 2007: 27), a steady state implies a constant physical stock of people and artefacts. It is useful here to distinguish between *growth*, meaning physical expansion, and *development* which can be non-physical. In natural systems, periods of growth always comes to an end (as for a human being or a natural forest) while dynamic development continues, including individual cycles of growth and decay. It is therefore self-deceptive and un-natural to anticipate an ever-lasting expansion, as modern economies appear to require.

Having recognised the need to reach a steady-state economy, the next question is to identify the appropriate size of the economy. The *maximum* possible stock of people and artefacts which can be maintained depends upon the carrying capacity of the environment and the efficiency with which energy and resources are used. But the *desirable*, or *optimum* level, will no doubt be considerably lower than this, since the ecological sacrifices required for maintaining the maximum possible stock are likely to be both socially and psychologically intolerable.

A number of studies have suggested that the present biophysical throughput of the global economy is exceeding the carrying capacity of the Earth. For example, estimates of the 'ecological footprint' of the global economy suggest that we are exceeding the planet's carrying capacity by around a quarter, which is clearly unsustainable (Wackernagel et al., 2005). Comparable indicators developed by the Wuppertal Institute suggest that we are currently using natural resources at twice the sustainable rate (Schmidt-Bleek, 2000).

Globally there is enormous inequality in the consumption of natural resources, with 20 per cent of the world's population accounting for about 80 per cent of the world's resource use. This suggests the need for major reductions in resource consumption in the affluent countries. For example, if a global target to halve resource use and environmental impacts were to be combined with a principle of equal per capita rights to use nature, European countries would need to reduce their load on nature by approximately a factor of ten (Schmidt-Bleek, 2000).

Absolute savings, not just efficiencies

Often targets for energy savings are expressed in terms of energy efficiency, which is a relative term – for example, kilometres travelled per litre of fuel. For a whole nation, energy efficiency is typically expressed as GDP per unit of energy consumption. But from an environmental point of view, it makes little sense to set overall environmental goals in terms of efficiencies, since improvements in efficiencies can easily go hand-in-hand with growing energy consumption. Impacts on the natural environment are only affected by *how much* energy we consume, and no credit is offered to getting more output or GDP out of each unit of energy consumption.

One illustrative and striking example of how difficult it is to distinguish between energy efficiency and energy savings – and to choose between the two – is found in the *Factor 10 Manifesto* mentioned above (Schmidt-Bleek, 2000). In some paragraphs, the authors stress that the factor 10 target refers to the absolute use of nature, while in other paragraphs the target is described as a factor 10 increase in efficiency – meaning a reduction in material input per unit of service. From an environmental perspective, only the former makes sense. A factor 10 target on efficiency would allow a growing economy to increase material use infinitely

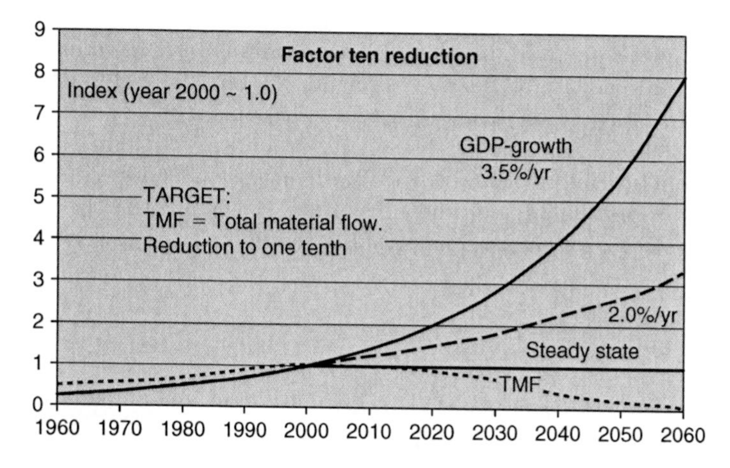

Figure 10.1 Factor 10 reduction

Note: With 3.5% annual growth in GDP, a factor 10 reduction in the absolute use of nature (TMF) by 2060 requires a factor 80 improvement in eco-efficiency compared to 2000. With a 2% annual growth in GDP, the factor would be 35 and in a steady state economy, just factor 10.

(see Figure 10.1). Such confusions and contradictions are common in the sustainability debate.

Figure 10.1 illustrates the dilemma growth economists are facing, when confronted with a requirement to reduce the absolute use of nature by a factor of 10, while at the same time maintaining a GDP growth rate of say 3.5 per cent per year. Within a few decades, *eco-efficiency* would need to increase by a factor of 80.

Rather than setting targets for a reduction in energy use relative to some anticipated and highly uncertain growth in output or consumption, targets should be expressed instead in terms of *absolute* energy consumption (Nørgård, 2001; Herring, 2006b), as is now increasingly common in carbon dioxide emissions policy. A change in terminology is also required. Soon after the energy crises of the 1970s, the terms *energy conservation* or *energy savings* were widely used by environmentally engaged politicians, environmentalists and researchers. But these were later replaced by the misleading term *energy efficiency*, since this was more acceptable to conventional economics and established interests (see for example Chapter 9). Denmark is one of the few countries where the term 'energy savings' has survived, even in policy debates.

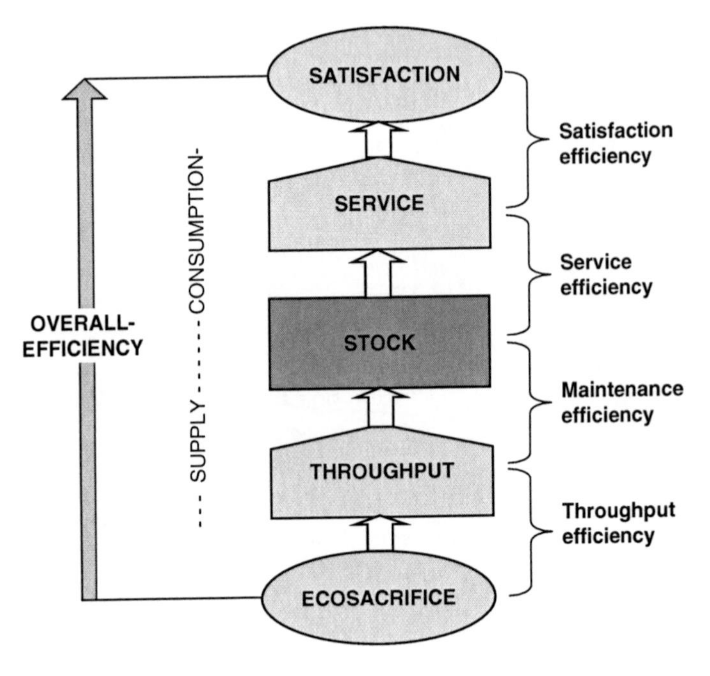

Figure 10.2 The whole economy

Note: The economy is here illustrated by a chain converting the means, ecological sacrifice, into the end, human satisfaction. Stock can be owned by either consumers or suppliers. The diagram can be applied to just energy or to the whole economy.

Whole economic system chain

Figure 10.2 shows a flow diagram of the whole economy, inspired by a model by Herman Daly (1996: 69), but here unfolded beyond services. This illustrates how the ultimate means – 'ecological sacrifice' – is converted into the ultimate ends, defined here as happiness, or human satisfaction. The diagram allows four different efficiency measures to be defined, including 'satisfaction efficiency' representing the ratio of human satisfaction to the level of services available (Nørgård, 2006a). The product of these gives the 'overall efficiency' of the economy.

For a sustainable economy, a steady-state stock of material artefacts needs to be maintained with a minimal throughput of resources. A reduction in this throughput can come about in three ways:

1. *Forced* upon the *supply* side when the environment cannot sustain sufficient throughput to maintain the stock, causing the economy to shrink or collapse.

2. *Voluntarily* chosen on the *consumption* side, as when people feel they have sufficient services and seek satisfaction in other ways, such as more free time. This should throttle the economy's throughput.
3. *Merging* the two limiting factors, as when it is collectively recognised that the benefits from increasing consumption are more than offset by the costs in terms of both work effort and eco-sacrifice. Trade-offs such as these are a major ethical issue in the quest for sustainable development.

The collective and voluntary choice of such an approach, based upon the notion of 'sufficiency', is the assumption behind this chapter. The alternative is considered to be the collapse of the world economy as a consequence of resource depletion and environmental pollution.

It is notable that the two extreme links in the chain, *satisfaction* and *eco-sacrifice*, are both difficult to measure. Nevertheless, they are the most essential elements in analysing the options for a humane and sustainable development. As Albert Einstein observed: *'Not everything that can be counted counts, and not everything that counts can be counted.'* The inadequacies of conventional measures of GDP in capturing both benefits (satisfaction) and costs (eco-sacrifice) have led several researchers to seek better measures of welfare and progress, discussed below.

Modifying gross domestic product

A variety of indicators have been developed based upon modifications to conventional measures of GDP. One is the Genuine Progress Indicator (GPI) which adjusts GDP to reflect environmental costs, income distribution, and the benefit of leisure time. According to the GPI, people in the United States are no better off today with respect to real economic welfare than in 1970, despite per capita GDP more than doubling over this period (Talberth et al., 2006). While there have been no GPI studies of the Nordic countries, the comparable Index of Sustainable Economic Welfare (ISEW) shows very similar trends for Sweden (Jackson and Stymme, 1996).

Both the GPI and the ISEW measure welfare in monetary terms and therefore do not break out of the straitjacket in the middle of Figure 10.2, to include the ultimate ends and means. But in recent decades, a number of researchers have begun to investigate the relationship between GDP and people's satisfaction or happiness (Layard, 2006; OECD, 2006). These studies include time series in single countries as well as comparisons of different countries. Very roughly, the studies conclude that beyond a certain minimum level, people are not made better off by higher incomes and increased consumption (Layard 2006: 32). As an example, the number of Americans who report being 'very happy' has remained relatively constant since 1950 (around one-third), although real income per capita has almost tripled over this period. The picture is similar in other Western countries and fits with the results of the GPI above.

Even the Organization for Economic Cooperation and Development (OECD) has begun to doubt the value of using GDP as a measure of welfare. One chapter of a recent report examined various indicators of happiness and life satisfaction in member countries and concluded that they were only weakly related to GDP. Noting that well-being is a more fundamental goal than GDP, the report observed that: 'It would be perverse to strive for faster growth in output if this entailed reducing the well-being of current and future generations' (OECD, 2006: 130). Nevertheless, the title of the report is still *Going for Growth*!

Overall economic efficiencies

While some of the alternative economic indicators like the GPI include measures of human welfare and environmental costs, few have attempted to measure the *overall efficiency* of the economy, as indicated in Figure 10.2. A measure of overall efficiency would be the ratio of some measure of satisfaction, or happiness, to some measure of ultimate cost in terms of ecological sacrifice. The Happy Planet Index, HPI (Marks et al., 2006) is such an overall efficiency indicator, with ecological footprints as the denominator (the cost) and the product of life expectancy and an index for the more subjective human satisfaction as the numerator (the benefit). The results from the HPI in different countries are often counter-intuitive. For example, a ranking of 178 countries places most Western countries near the bottom, with the US as number 150, France 129, Canada 111, UK 108, Japan 95 and Germany 81 – mostly due to these countries' high ecological footprint. In general the Central American countries score the highest, because they manage to combine a long life expectancy and high life satisfaction with a low ecological footprint. The clear implication is that a high well-being does not necessarily require a high level of consumption, just as high consumption does not guarantee high well-being (Jackson, 2005; Marks et al., 2006). A more consistent result is that high consumption invariably leads to high footprints.

No doubt, the transition to a sustainable and steady-state society will require significant economic and social changes: 'a modification of society comparable in scale to only two other changes: the Agricultural Revolution of the late Neolithic and the Industrial Revolution of the past two centuries' (Meadows et al., 2004: 265). Social values will need to change somewhat, both to bring about this transition and to adapt to the new situation (Christensen and Nørgård, 1976). Nevertheless the transition need not lead to a net human sacrifice – on the contrary (Jalas, 2002; Jackson, 2005; Sanne, 2007). While the cost of achieving a sustainable economy is often calculated to be substantial in narrow, monetary terms, once human satisfaction and eco-sacrifice are considered (Figure 10.2), the balance appears quite different.

The above-mentioned attempts to quantify overall efficiency should not obscure the fact that the path of development needs to be decided through open dialogue and democracy, rather than dictated by what happens to be easily quantifiable. As argued in the next section, it should not be difficult to achieve a sustainable economy with a net gain in human welfare. This derives from a number of sources, but in particular from a rebalancing between consumption and leisure.

Sufficiency and less work

Economic satiation seldom implies that people don't want any more improvement in their lives. It just indicates that their development has reached a level where consumption is sufficient, and their marginal demand is outside the monetary frame (Johansson, 2007). The most obvious example of such non-economic human welfare is reduced work and increased leisure time, which furthermore is a key to lowering environmental impacts (Sanne, 2000).

Work-time reduction revived

Working time has historically been a variable production factor, although a sluggish one, with changes taking decades or even centuries. It reached a peak during industrialisation in the 1800s when more than 70 hours of hard work per week, or 3500 hours per year, were common (Sanne, 1995: 145; Beder, 2000; Schor, 2005). Since then, average work time has declined to around half, and in the 1950s the future was commonly envisioned as a leisure society with for instance only fourteen hours' work per week, as suggested by a US Senate subcommittee (Honoré, 2004: 163). Such visions have not materialised however. Despite continuous technological productivity gains and substantial increases in affluence, the decline in work time per employee in Western Europe came to an end after 1980 and in 2000 was around 1500–1600 hours per year (Schor, 2005). In the Nordic countries, paid work hours per working-age person are essentially the same as thirty years ago (Schor, 2005), and working

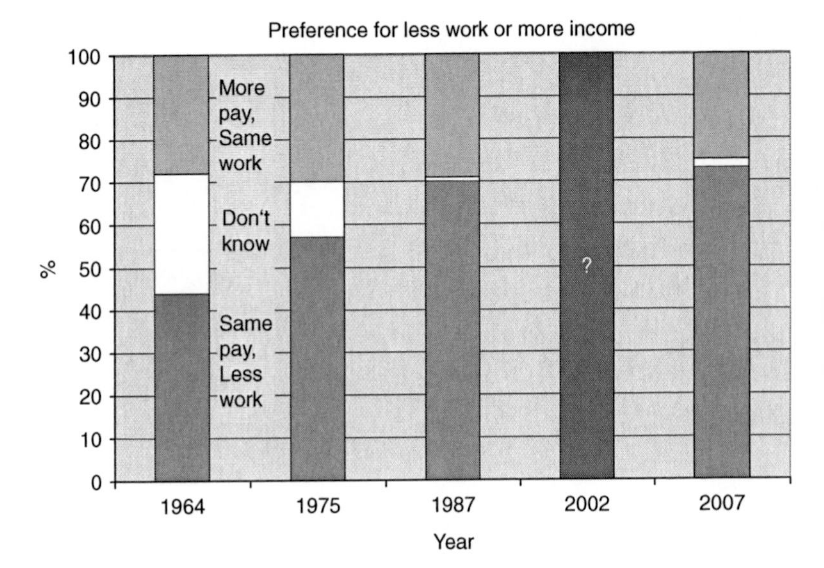

Figure 10.3 The choice between more income and reduced work time

Note: Sociological surveys over 43 years show for Denmark an increased interest in turning productivity increase into more leisure rather than more consumption (Platz, 1988 and IFKA 2007).

hours *per family* have greatly increased as a result of women entering the workforce (Sanne 1995: 42).

For politicians pursuing high GDP growth, the situation in affluent countries can seem challenging. With no growth in population and nearly all women in the labour force, growth can be obtained only from labour productivity gains, which can be hard to maintain at 2 per cent per year. For politicians aiming at a steady-state economy, however, the outlook is much more positive, especially if the drive towards less work time can be revived. The productivity gains from improved technology, including that from energy efficiency, could then be used to slow down instead of consuming more goods and services. Fortunately, this sustainability quest appears to fit well with trends in public preferences.

Since 1964, the National Institute of Social Research in Denmark has conducted extensive surveys of how Danes use their time, and how they would like to use it (Platz, 1988; Körmendi, 1990).

One of the questions concerns the choice between more income and reduced work time. The results are shown in Figure 10.3. The fraction preferring less work appears to have grown over time, reaching 70 per cent in 1987. In their next survey in 2002 the National Institute of Social Research for unclear reasons left out just this question. Fortunately, in 2007, another institute, IFKA (2007), took up the question and showed a continuation of the trend, now reaching 73 per cent preferring less work.

Similar trends towards consumption saturation are found in other affluent countries, and particularly in the Nordic countries (Sanne, 1995: 53, 2007: 49). The attitudes indicated by such surveys are unsettling to most politicians, who tend to be primarily concerned with increasing production and consumption. Similarly, these results are usually ignored by labour unions, even though they have historically fought for shorter work time.

Why don't people work less, if they want to? The answer is partly that few employers offer such choice, and partly social pressures (Galbraith, 1973: 236; Sanne, 1995: 74). The work market is not free and usually involves a choice between 35–40 hours per week or zero. Also, the quest for equity and solidarity in sustainable development calls for collective agreements on work time.

Spending leisure time

A common misunderstanding of the above survey results is that with more leisure people will consume more. This, of course, is not possible. If people choose to have more leisure *instead* of more income, their level of consumption will remain constant. Whether more leisure time will result in higher *energy* consumption obviously depends on how the time is spent. For example, car driving has one of the highest rates of energy consumption per hour, about five litres of petrol or 50 kWh, while in contrast, reading only consumes around 1 kWh (Jalas, 2002). For comparison, an hour less at work in Denmark is estimated to save an average of roughly 25 kWh of energy consumption, derived from the total energy consumption for workplaces (Danish Energy Authority, 2007) divided by the total volume of working hours in the country (Statistics Denmark, 2007).

Extra leisure time can be spent not only with low energy intensity, but even with a 'negative' intensity as when net reductions in energy consumption are obtained from slowing down. For example, reducing car speeds from 130 km to 80 km per hour not only saves around half of the fuel used per km, but considering slower car driving as a way to spend extra free time, it saves about two-thirds of the fuel per hour. Choosing to spend some of the extra free time to replace commuting by car with walking and cycling, possibly combined with public transport, is an option with significant 'negative' energy consumption – as well as improved health and well-being (Nørgård, 2005). Not surprisingly, since the *rate* of material throughput is the primary source of environmental impacts (Figure 10.2), 'slowing down' in general may be considered a necessary strategy to achieve sustainability.

Equity and other sufficiency drivers

The end goal of satisfaction and the trends in people's aspirations shown (see Figures 10.2 and 10.3) suggests that *less work* rather than more consumption is a key to higher satisfaction. Even though preferences for less work may not be guided by moral wishes to preserve the environment, a trend towards voluntarily desisting from increasing consumption is environmentally beneficial and should be encouraged and supported by policy.

Higher equity may play a double role in the quest for sustainable development. First, the recognition of a limited world will tend to make demands for equal rights to the use of natural resources more morally and politically legitimate. Secondly, in affluent countries the growth in consumption appears to be primarily driven by differences in *relative* income (Keynes, 1931; Jespersen, 2004). Consequently, high equity in affluent economies tends to hamper consumption growth – that is, promote economic satiation – as suggested in a warning to Denmark from the OECD (1982). Compared to other regions, Nordic countries have been characterised by relatively high equity and public welfare, financed by high and progressive income taxes that have the additional advantage of discouraging people from becoming too addicted to work (Layard, 2006: 155). Global redistribution of wealth could in part be achieved through voluntary means, such through fair trade schemes which pay producers in developing countries higher prices than what market forces dictate.

Another driver of limiting consumption and working less in the affluent countries could be health considerations. As the economist J. K. Galbraith noted: 'Virtually all of the increase in health hazards is the result of increased consumption' (Galbraith, 1973: 279). Support for this view can be found, for instance, in the obesity problem in the US. This causes 400,000 premature deaths annually and in the view of a US minister of health: 'is definitely one side effect of getting wealthier' (Samuelson, 2004). Increased wealth tempts people to use cars excessively and to eat too much food (Nørgård, 2005).

Sharing work more equally is a key issue. Recent research in happiness reveals that being unemployed is a significant source of misery (Layard, 2006: 67). On the other hand, working fewer hours seems to be highly desirable (Figure 10.3). All this new investigation into happiness and satisfaction has the potential to direct attention towards non-economic welfare options, rather than simply focusing on maximising income.

So far we have not discussed reducing consumption, only ceasing its growth. But already, more than 20 per cent of employed people in Denmark would prefer to work less even with a corresponding decline in income (Bonke, 2002: 51). The general low and declining preference for more consumption is even more remarkable in light of the continuing commercial and political pressure in the opposite direction with longer shopping hours, electronic payments, increased advertisement budgets, globalisation, etc., not to mention the ongoing 'selling of the work ethic' (Beder, 2000). Considering this pressure to work and consume more, it could be argued that people in affluent countries are not only saturated but effectively 'super-saturated' in terms of consumption. Once the option for working less is let loose and encouraged as a contribution to achieving sustainability, the balance between work time and consumption might settle at a new and lower level. Having more time free from paid work could in a kind of virtuous circle reduce the need for paid child care, for cars, for private deep freezers and so on, and thereby further reduce the need for income and work time.

What is discussed here as reducing work time, for instance as hours per year, should be interpreted broadly as work input into production. In some tasks, it could take the form of the same (or even longer) working time, but with a lower productivity, provided this led to a higher degree of human satisfaction, better health and personal development (Nørgård, 2006b). A reduction in labour productivity can reduce energy consumption per economic output.

Sufficiency and the productivity elasticity of leisure

The income elasticity of a consumer good is the percentage increase in consumption of that good following a percentage increase in income. Estimates of income elasticities have been used by number of authors to quantify the rebound effect from savings on energy bills (Nässén and Holmberg, 2007).

Aggregate increases in income normally result from improvements in labour productivity (output per hour worked). But as we have seen, people could potentially benefit from productivity increases in other ways – notably by working fewer hours. If, for instance, one-tenth of the annual productivity gains was turned into reduced working time, while 90 per cent ended up as higher income (and hence consumption), leisure could be said to have a 'productivity elasticity' of 0.1 while total consumption had a 'productivity elasticity' of 0.9. Such values might approximately reflect the experience of the past fifty years in Europe, but may not necessarily reflect the patterns people would have preferred. If the choice between income and increased leisure better reflected the preferences suggested by Figure 10.3, the productivity elasticity of leisure would increase. This elasticity could then be included when estimating the rebound effects from productivity improvements. In a 'fully satiated' economy, the productivity elasticity of leisure would be unity, meaning that all increases in productivity would end up as increased leisure time, with no net increase in production, real income and hence consumption.

Concluding reflections

Where basic needs are not satisfied, it is reasonable for economic growth to take priority over environmental benefits and energy savings. In these circumstances, the rebound effect should be beneficial for human welfare, since it speeds up economic development. And when the economy has reached a state of collective sufficiency, such as outlined in this chapter, there may be no rebound effect, since all productivity gains (including those from energy efficiency) should be turned into forms of welfare beyond the money economy, such as increased leisure. In these circumstances, the full environmental benefits of technological efficiency gains should be realised. Hence the rebound effect is only a problem when we argue for sustainability but dare not face the fundamental issue: our mania for economic growth.

Political and economic thinking in a sufficiency based steady-state economy will be very different from that at present. Fortunately, in affluent countries there are some signs that public attitudes may encourage a shift towards sufficiency. These include preferences for exchanging consumption growth for less work time and a willingness to have fewer children to ease population pressures. These changes can thus create opportunities for a gradual democratic transition towards sustainability, but so far these options are not on the political agenda.

Policies to pursue a sustainable steady-state economy based on sufficiency may frequently be the opposite of current policies, which aim primarily at increasing GDP and aggregate consumption. Socially, a sustainable economy will no doubt require more equal sharing of income as well as of work to prevent unemployment and poverty.

In discussing visions for the future, it is important to take a holistic approach, instead of expecting significant changes by just dropping one measure like energy efficiency into an otherwise unchanged society. In our present endeavour towards a sustainable future, the problem is not that we strive for higher technical energy efficiency. The problem arises from *only* striving for such higher efficiency, and in a technological ecstasy leaving morality and sufficiency behind.

Acknowledgements

The topic of my chapter in this book takes up an issue which for long has been close to my heart, so I was happy and flattered to be asked to contribute. I am very grateful to the editors, Horace Herring and Steve Sorrell, who have provided valuable and constructive ideas and comments on my manuscript, but as far as the content is concerned I do of course take full responsibility.

References

Beder, S. (2000) *Selling the Work Ethic: From Puritan Pulpit to Corporate PR*, Melbourne: Scribe.

Bonke, J. (2002) *Tid og velfærd* (Time and Welfare), Copenhagen: Danish National Institute of Social Research, Report 02:26.

Brookes, L. (1990) 'The greenhouse effect: the fallacies in the energy efficiency solution', *Energy Policy*, 18(2): 199–201.

Christensen, B. L. and J. S. Nørgård (1976) 'Social values and the limits to growth', *Technological Forecasting and Social Change*, 9: 411–23.

Daly, H. E. (1977) *Steady State Economics*, San Francisco: W. H. Freeman and Company.

Daly, H. E. (1990) 'Towards some operational principles of sustainable development', *Ecological Economics*, 2: 1–6.

Daly, H. E. (1996) *Beyond Growth: the Economics of Sustainable Development*, Boston: Beacon Press.

Daly, H. E. (2007) *Ecological Economics and Sustainable Development, Selected Essays of Herman Daly*, Cheltenham: Edgar Elgar.

Danish Energy Authority (2007) *Facts and Figures*, www.energistyrelsen.dk

Galbraith, J. K. (1973) *Economics and the Public Purpose*, Boston: Houghton Mifflin Company.

Herring, H. (2006a) 'Energy efficiency: a critical view', *Energy*, 31(1): 10–20.

Herring, H. (2006b) 'Confronting Jevons' Paradox: does promoting energy efficiency save energy?', *IAEE Newsletter*, Fourth Quarter, pp. 14–15.

Hicks, J. R. (1948) *Value and Capital*, 2nd edition, Oxford: Oxford University Press.

Honoré, C. (2004) *In Praise of Slow: How a Worldwide Movement is Challenging the Cult of Speed*, London: Orion Books Ltd.

IFKA (2007) *IFKA's årbog: Danskerne 2008* (IFKA's yearbook: *The Danes in 2008*). Copenhagen: IFKA (Institute for Business Cycle Analysis), pp. 37–40, www.ifka.dk

Jackson, T. (2005) 'Live better by consuming less? Is there a "double dividend" in sustainable consumption?', *Journal of Industrial Ecology*, 9(1–2): 19–36.

Jackson, T. and S. Stymme (1996) *Sustainable Economic Welfare in Sweden: a Pilot Index 1950–1992*, Stockholm Environmental Institute.

Jalas, M. (2002) 'A time use perspective on the materials intensity of consumption', *Ecological Economics*, 41(1): 109–23.

Jespersen, J. (2004) 'Sustainable development and macroeconomic stability', in L. Reich and I. Røpke (eds), *The Ecological Economics of Consumption*, Cheltenham: Edward Elgar.

Johansson, B. (ed.) (2007) *Konsumera mera – dyrköpt lycka* (Consume More – Dearly Bought Happiness), Stockholm: Formas publisher, www.formasfokuserar.se

Keynes, J. M. (1931) *Essays in Persuasion*, London: Macmillan.

Körmendi, E. (1990) 'Preferences and time use', in G. V. Mogensen (ed.), *Time and Consumption*, Copenhagen: Statistics Demark, pp. 143–61.

Layard, R. (2006) *Happiness: Lessons from a New Science*, London: Penguin.

Marks, N., S. Abdallah, A. Simms and S. Thompson (2006) *The Happy Planet Index: an Index of Human Well-being and Environmental Impact*, London: New Economic Foundation.

Meadows, D. H., D. L. Meadows, J. Randers and W. W. Behrens III (1972) *The Limits to Growth*, New York: Universe Books.

Meadows, D. H., J. Randers and D. L. Meadows (2004) *The Limits to Growth: the 30-Year Update*, White River Junction, Vermont, USA: Chelsea Green Publishing Company.

Mill, J. S. (1900) *Principles of Political Economy* (original version published in 1848). Revised edition, Vol. II, New York: Colonial Press.

Nässén, J. and J. Holmberg (2007) 'Quantifying the rebound effect of energy efficiency and energy conserving behaviour in Sweden', submitted to *Energy Efficiency*.

Nielsen, J. S. (2006) 'The green revolution – now also at the right wing' (in Danish), Interview in newspaper *Information*, 2–3 September, Copenhagen.

Nørgård, J. S. (1974) *Technological and Social Measures to Conserve Energy*, Dartmouth College, New Hampshire: Thayer School of Engineering, Report DSD 26 (available from the author).

Nørgård, J. S. (2000) 'Danish energy saving policy: the past, the present, and how it should be', Proceedings of Electricity for a Sustainable Urban Development Conference, Lisbon 2000, Paris: Union Internationale pour les application de l'Electricité, pp. 365–80.

Nørgård, J. S. (2001) 'Can energy saving policy survive in a market economy?', Proceedings of the European Council for an Energy Efficient Economy, Summer Study, Paris: ADEME, Volume I, pp. 261–73.

Nørgård, J. S. (2005) 'Under-use of body energy and over-use of external energy', Proceedings of the European Council for an Energy Efficient Economy, Summer Study, Stockholm: ECEEE, pp. 243–52.

Nørgård, J. S. (2006a) 'Consumer efficiency in conflict with GDP growth', *Ecological Economics*, 57: 15–29.

Nørgård, J. S.(2006b) 'Limiting labor input is an overall prerequisite for sustainability', Proceedings of the International Society for Ecological Economics, New Delhi Conference, 16–18 December (available from the author).

Nørgård, J. S., B. Brange, T. Guldbrandsen and P. Karbo (2007) 'Turning the appliance market around towards A++', Proceedings of the European Council for an Energy Efficient Economy, Summer Study, Stockholm: ECEEE, pp. 155–64.

OECD (1982) *Economic Surveys 1982–83: Denmark*, Paris: Organization for Economic Co-operation and Development.

OECD (2006) *Economic Policy Reforms: Going for Growth 2006*, Paris: Organization for Economic Co-operation and Development.

Platz, M. (1988) *Arbejdstid 1987* (Working Time), Copenhagen: Danish National Institute of Social Research.
Samuelson, R. J. (2004) 'The afflictions of affluence', *Newsweek*, 22 March.
Sanne, C. (1995) *Arbetets tid – Om arbetstidsreformer och konsumption i välfärdsstaten* (Working Hours in the Age of Work), Stockholm: Carlsson Bokförlag.
Sanne, C. (2000) 'Dealing with environmental savings in a dynamic economy: how to stop chasing your tail in the pursuit of sustainability', *Energy Policy*, 28(6–7): 487–95.
Sanne, C. (2007) *Keynes barnbarn – en bätre framtid med arbejde och välfärd* (Keynes' Grandchildren – a Better Future with Work and Welfare), Stockholm: The Swedish Research Council, www.formas.se
Schmidt-Bleek, F. (2000) *Factor 10 Manifesto*, Wuppertal Institute, www.factor10-institute.org
Schor, J. B. (2005) 'Sustainable consumption and work reduction', *Journal of Industrial Ecology*, 9(1–2): 37–50.
Simmons, M. R. (2000) 'Revisiting *The Limits to Growth*: could the Club of Rome have been correct, after all?', *Energy Bulletin*, 30 September, www.energybulletin.net/1512.html
Statistics Denmark (2007) *Working Time Account*, Statistics Denmark, www.danmarksstatistik.dk
Talberth, J., C. Cobb and N. Slattery (2006) *The Genuine Progress Indicator 2006: a Tool for Sustainable Development*, California: Redefining Progress, www.rprogress.org
Tinbergen, J. and R. Hueting (1991) Chapter 4 in R. Goodland, H. Daly and S. Serafy (eds), *Environmentally Sustainable Economic Development: Building on Brundtland*, Washington, DC: Environment Department, The World Bank.
Vermes, T. (1990) 'Frihandel er en fare for jordas miljø' (Free trade is a risk to the global environment), interview with Haavelmoe in the newspaper *Klassekampen*, Oslo, 3 November.
Wackernagel, M. et al. (2005) *Europe 2005: the Ecological Footprint*, Brussels: WWF European Policy Office, www.footprintnetwork.org
WCED (1987) *Our Common Future*, World Commission on Environment and Development, Oxford: Oxford University Press.

11
Sufficiency and the Rebound Effect

Horace Herring

In this book so far there has been little if any discussion of why consumers might want to invest in energy efficiency. The most obvious, and common, reason is to save money which, after repaying the costs of investment, can then be spent on other activities, all of which use energy to some extent. Hence the rebound is never zero. But this chapter is not going to discuss the economic reasons for investment by consumers, but instead will focus on the non-economic, chiefly environmental and ethical, reasons for pursuing energy efficiency.

For most environmentalists, and echoed by government adverts, energy efficiency is a way to save the world – from global warming, from energy shortages, and from regional conflicts over oil and gas resources. In this view improvements in energy efficiency lead to an absolute reduction in energy consumption, or (less explicitly stated) reduce the rate of growth of energy consumption and thus lessen global problems (from what they would have been without energy efficiency). This view is integral to the concept of sustainable development with its three strategies of lowering consumption by consuming more efficiently, differently and less.

Previous chapters in this book have challenged the idea that greater efficiency leads to lower consumption, so this chapter concentrates on the ideas of reducing consumption through consuming differently and consuming less. Both of these explicitly question our reasons for consumption. This can lead to a chain of questions and answers, such as: Why do we use electricity? Because we want electric light. Why do we want light? To feel secure and comfortable. Why do we want to feel secure? Because we are afraid of the dark. And why are we afraid of the dark? Hence the solution to this ultimate cause – fear of the dark – need not be more electric lighting, but instead education (to banish superstition), more police patrolling (to reassure the public), medical help (psychiatrists to rid us of our phobias), or different forms of entertainment (TV rather than reading) that use lower levels of lighting. This example may seem absurd for the individual, but may be very valid for public services. For instance, is more and better street lighting an answer to rising levels of crime? Some even advocate less street lighting, so we can once again appreciate the stars!

So what role for energy efficiency when we want to consume differently and less? The strategy of consuming differently is based on the proposition that services consume fewer resources than products. Hence, it is argued, home deliveries of internet shopping or take-away meals could save energy (compared to going to the shops yourself). There is already an extensive literature on this strategy (e.g. Roy, 2000; Halme, 2005), so instead we will concentrate on the third strategy of consuming less, a policy called 'sufficiency', 'voluntary simplicity' or 'downshifting'.

The chapter first provides a perspective on why we consume, then examines the ideas behind sufficiency, questions its effectiveness especially if it became widespread, and finally asks whether it is a feasible goal that governments should support under the banner of sustainable development.

Why do we consume?

Before I write more about these three energy strategies, I would like to pose the question 'why do we consume energy?' And here I am going to take an anthropological approach to the question of consumption, which is influenced by the works of Mary Douglas and her colleagues, and subsequent 'energy anthropology' writers like Hal Wilhite, and also the socio-cultural approach of Elizabeth Shove (Douglas and Isherwood, 1979; Shove, 2004; Wilhite, 2005). So why do we consume? As Mary Douglas observed:

> A person wants goods for fulfilling personal commitments. Commodities do not satisfy desire: they are only the tools or instruments for satisfying it. Goods are not ends. Goods are for distributing, sharing, consuming or destroying publicly in one way or another. To focus exclusively on how persons relate to objects can never illuminate desire. Instead research should focus on the patterns of alliance and authority that are made and marked in all human societies by the circulation of goods. Demand for goods is a chart of social commitments, graded and timetabled for the year, or the decade, or the lifetime ...

> The main objective of consumption is the desired pattern of social relations. The material objects only play an ancillary role; goods are battle standards; they draw the line between good and evil; and there are no neutral objects.
>
> (Douglas et al., 1998: 202)

This is a completely different interpretation from the standard economic model of people as rational profit (or utility) maximising individuals. Rather it portrays people as social beings located in a particular culture. As many energy authors have pointed out, people want to consume energy services rather than energy carriers (e.g. Patterson, 1990). That is, we want better heating, hot water, refrigeration or lighting rather than more electricity and gas. But few authors ask why we want energy services; what underlies our desire for heating or hot water? One who has is Elizabeth Shove, and she groups the ultimate purpose of household energy services into the trio of comfort, cleanliness and convenience which are not only physical factors, but also psychological and ethical ones (Shove, 2004). Comfort and convenience can be seen as individual, materialist values, heavily promoted by a consumer culture. But cleanliness is more a cultural or social value, and plays a major role in religious teachings, with their notions of purity in food and behaviour, especially sexual. It also plays a major role in environmental thinking, with its concept of man-made pollution versus natural purity. Hence the relevance of Mary Douglas' famous work *Purity and Danger: an Analysis of Concepts of Pollution and Taboo* (1966).

The worst insult you can level against people is that they are 'dirty', and this is, and has been, used to justify discrimination and persecution of social and ethnic groups. It is also heavily marketed by energy companies: clean electricity versus dirty coal, or by environmentalists: clean solar against dirty nuclear. So the concept of cleanliness (and pollution) is a key battleground in the public acceptance of energy sources.

And, going further we can ask why we want more comfort or convenience? Is it to improve our 'standard of living' or 'quality of life'? The former is arguably a more objective concept since it can be measured in physical or monetary terms, while the latter appears more subjective and is heavily influenced by environmental and ethical concerns. And above this desire for quality of life may be ethical or religious concerns. This hierarchy of motives is shown in Figure 11.1.

This suggests that attempts to influence or change our demand for energy would benefit from improved knowledge of people's intermediate and ultimate concerns. For example, do more efficient light bulbs fit

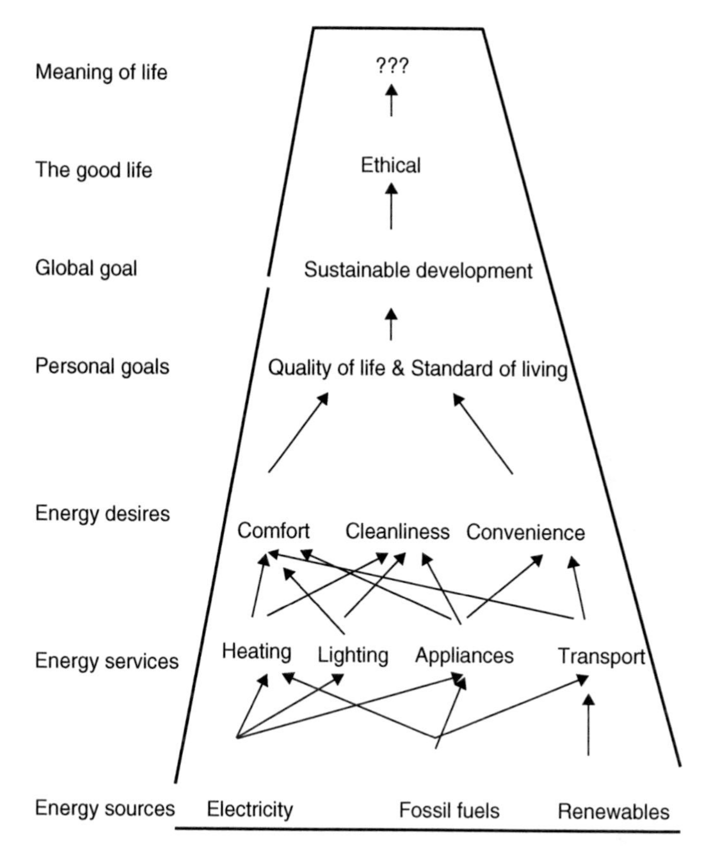

Figure 11.1 The hierarchy of motives: from energy sources to the meaning of life

in with people's desire for greater convenience? Do solar water heaters tally with people's desire for greater cleanliness? Do government adverts advising people to turn down the thermostat by 1°C resonate with our desire for comfort? Do appeals to consume less strike a chord with people wanting to improve their quality of life?

Once this hierarchy of motives is accepted, then paradoxical energy-efficiency actions can be more easily understood. Installing double-glazing has been far more popular than cavity wall or loft insulation despite its higher cost and worse economic return. This can be explained by the (perceived) comfort and cleanliness it brings, the convenience of installation, and its visible social prestige – another source of comfort. Again with solar water heating, an important factor in its installation is its (perceived) cleanliness: that is, the hot water comes from solar energy which is considered as a non-polluting and green energy source compared to electricity or gas (Caird and Roy, 2007).

In Chapter 8 we saw the importance of time savings and their impact on energy use, so it is no surprise that consuming differently will be very popular if it is more convenient. Hence the popularity of internet shopping, which not only saves time (very convenient), but can be a more comfortable shopping experience besides being considered as a more green way to shop. Seldom considered by consumers are the vast amounts of resources used by IT networks and computer equipment, and the transport energy needed to deliver goods (Rejeski, 2003). Finally, how does consuming less fit in with our hierarchy? Is less consumption more comfortable, convenient or clean? How does it fit in with an improved quality of life? Is consuming less an ethical or green way of life?

Defining less consumption

Most people, especially if they are poor, want to consume more but are constrained by their incomes – they cannot afford the goods and services they desire. The wealthier people are the more goods and services they consume, so one fairly obvious way to lower consumption (and the associated environmental impacts) is to reduce income (Hertwich, 2006). A number of authors have recommended that people *voluntarily* reduce their incomes, consume fewer goods and services and thereby develop 'greener' lifestyles that have lower environmental impacts (Trainer, 1985; Ghazi and Jones, 2004; Princen, 2005; Petherick, 2007). This so-called 'sufficiency strategy' is generally assumed to rely upon individual action:

> Two concepts are needed to define sufficiency behaviour. First, it presupposes *purchasing power*: Those who are to alter their behaviour towards less consumption must be *able to* consume. Their purchasing power either remains unused or is itself reduced through working and earning less. The second concept is *environmental motivation*: We all limit consumption at some point, for many reasons. I am however confining the definition to the costs of non-consumption that are voluntarily traded for the benefits of believing one is relieving human pressure on planetary resources and thus benefiting other (present or future) humans or other species.
>
> (Alcott, 2008: 771)

If millions of people did this in rich countries, then, so the argument goes, there would be less consumption of energy resources, so less carbon emissions, and hence the world could avert catastrophic climate change. Moreover, a key component of this argument is that such choices can lead to happier and more fulfilling lives (see Chapter 10).

But are these arguments behind sufficiency valid? Can people voluntarily and permanently reduce their incomes? Is this likely to be a feasible strategy for more than a few individuals in rich countries? And even if it was, would it make any difference to global consumption and emissions? Is it an effective policy that governments can realistically encourage and what about its impact on national income and economic growth?

Income reduction

People's income can rise and fall for a wide range of reasons throughout their lives. Reductions in income as a result of unemployment, sickness, disability and redundancy are considered undesirable, even tragic, events and something to be avoided. Some falls are only temporary, such as career breaks for education and training, with the expectation that income will rise in the future. Others reflect domestic and family circumstances – such as a shift to part-time or less demanding work so as to care for young children or elderly relatives. Finally, some reflect job dissatisfaction and failure; better to quit now than be fired later. Just as there is the euphemism that a minister leaves his job to 'spend more time with his family' so the borderline worker can say she quits in order to 'spend more time in her organic garden'.

Likewise can those living on low incomes attempting to live a green lifestyle say they are practising sufficiency? Could they really be bond traders or corporate lawyers in the City earning millions? Does not our current income reflect our employment prospects? Thus by what criteria can we judge those who claim they have voluntarily chosen a lower income and less consumption for environmental reasons? Is it through choice or through circumstances? A survey of people practising voluntary simplicity in the Pacific north-west of the USA found that only a very small portion of those interested in it were motivated by altruistic or environmental reasons (Zavestoski, 2001). Most were attracted to it after an 'identity crisis' (caused by personal trauma such as divorce, death or children leaving home). They looked to it as source of meaning in their lives, with the hope of it alleviating the excessive stress, agitation, malaise and despair they were feeling in their consumer lifestyles. As Stephen Zavestoski (2001: 180) remarked, 'anti-consumerist sentiments seem to emerge out of a self-interested motivation to assert an authentic self', and thus the desire for sufficiency may well be more an emotional response to a personal crisis than a calm decision of how to respond to a global environmental problem.

Finally, I must stress that sufficiency is not the same as personal denial, abstinence and poverty (as practised by many religious orders). As the earlier quotation from Blake Alcott demonstrates, the goal of sufficiency is not personal, but global salvation. Thus it is more about politics than ethics, more to do with environmental campaigning than personal development. It can be viewed as a propaganda statement about the consumer society, and located in the long-standing tradition of using a consumer boycott as a tool to achieve change in society. This tactic dates back at least two centuries, to the boycott of West Indian sugar by the anti-slavery campaign. It was used widely in the twentieth century, for instance, by the civil rights movement against racist white-owned stores in the USA in the 1960s, and against South African products by the anti-apartheid campaign in the 1970s and 1980s.

Thus the call by green authors and environmental groups for people to practise sufficiency (and to publicise their efforts) should be seen more as a campaigning tool than a realistic effort to change global consumption. The call to refuse to be part of consumer society, to boycott its wares, and to live the simple life is a long-standing one and has so far been resoundingly unsuccessful (Shi, 1985). The global economy has continued to expand and with it the ethic of consumerism. What has changed is the efficiency and environmental impacts of individual goods and services – often in response to shifts in consumer demand and sometimes brought about by citizen and consumer campaigns. Not only can new technologies provide more comfort and convenience, but also more (environmental) cleanliness: less pollution in production and use. Thus achieving a 'greener' life is more a function of technology than changes in income levels.

As an example, the nineteenth-century call to boycott the dirty coal fire was rendered obsolete by the development and widespread use of twentieth-century gas central heating. Similarly the call to boycott dirty fossil fuels could be rendered obsolete by their replacement by twenty-first century renewable (or nuclear) energy sources. More income to buy what is clean may be more successful than less income to boycott what is dirty. Thus the call to sufficiency, to lower consumption voluntarily on environmental grounds, may make good copy for green groups in the campaign against global warming. It may also provide a temporary refuge for the casualties from the consumer society, but as a means to alter global consumption patterns it has a major obstacle: the rebound effect.

The sufficiency rebound effect

Whilst the impact of the rebound effect has been widely studied for energy efficiency, it has been mostly ignored for sufficiency. The first author to give extensive analysis to its impact on sufficiency is Blake Alcott (2008). Using arguments derived from the rebound debate on energy efficiency, he argues that the widespread impact of reducing demand for a commodity (like energy) would be to lower its global price, thereby encouraging greater consumption by (marginal) consumers. He remarks that 'some of what was "saved" through non-consumption is consumed after all – merely by others'; and that 'the sufficiency rebound then amounts to a passive, rather than intentional, transfer of purchasing power to marginal consumers' (Alcott, 2008: 775).

The extent of the rebound, and even the existence of backfire, will depend upon the responsiveness of demand and supply to changes in price: or more precisely, the relative slope of the demand and supply curves for the commodity in both rich and poor nations. A high rebound is likely to occur when the supply of the commodity is unresponsive to changes in price (i.e. the supply curve is flat) while the demand for the commodity is relatively responsive to changes in price (i.e. the demand curve is steep). A low rebound is likely to occur when the opposite is the case. In the case of energy, both supply and demand are relatively unresponsive to changes in price in the short term, but much more responsive in the long term. For example, given the low cost of oil production relative to the price of retail products, profits are likely to be high enough to withstand substantial price falls without curtailing supply. As Gary Duncan (2008) wrote recently in *The Times*:

> Moves by 'green consumers' and nations to cut energy use will prove pointless if oil-producing states simply maintain production even as prices fall, so that other countries more careless of the dangers of global warming then buy, and burn, all the fuel saved, enjoying an implicit subsidy from the West as they do so.

Rebound effects may be lower in circumstances where it is difficult for suppliers of consumer goods to find alternative markets or where there are superior alternatives. Take the decision to phase out incandescent light bulbs by the UK and other OECD governments. This boycott of bulbs will either force manufacturers to cease production or to find new markets in developing countries. However, no matter how cheap the bulbs, demand only exists where there is an electricity supply. There is also a technically superior, if more costly, alternative: the compact fluorescent light bulb (CFL). Whether this becomes the dominant product depends not only on selling price but also its profit margins for manufacturers and retailers and the rate of increase of electricity supply to the billions of the poor currently without electricity. Thus the decision to boycott incandescent light bulbs by rich countries could hasten their eventual disappearance from the world, to be replaced by a superior alternative – the more economical and longer lasting CFLs. Whether this would lead to a reduction in the amount of energy for lighting is another matter, given our intense desire for illumination (Fouquet and Pearson, 2006).

The sufficiency strategy not only implies earning and consuming less but also consuming differently. Sufficiency, of necessity, brings a change in lifestyle. Energy services will be met in different ways reflecting changed incomes and time availability: travel by bike and bus instead of by car, eating by cooking fresh food instead of take-away or frozen meals; washing and cleaning by solar water heaters instead of gas boilers; room heating with wood stoves instead of electric fires. There will be many choices to be made as to how best to satisfy our desires for comfort, convenience and cleanliness. These choices are currently being explored (and reported on) by sufficiency practitioners, and are only being slowly absorbed by mainstream society.

Economic impacts of sufficiency

One of the main arguments made against the sufficiency strategy is that it would lead to national economic recession and unemployment due to the reduction in consumption and the corresponding impact on economic growth. This is an important point. Most sufficiency advocates have failed to recognise the *structural dependence* of modern economies upon continued economic growth – and hence upon continued increases in consumption as measured by GDP. As a consequence of this, governments are unlikely to actively promote the sufficiency strategy – whatever its long-term benefits – if it threatens the short-term objective of continued economic growth.

The dependence of modern economies upon economic growth has a number of economic, social and cultural origins that are beyond the scope of this chapter. However, it is worth highlighting one particularly important factor, namely the nature of modern monetary systems.

The key problem is that around 97 per cent of the money in circulation is created not by governments but by commercial banks as interest-bearing debt:

> If a bank makes a loan, nothing is lent, for the simple reason that there is nothing of substance to lend. The bank makes what it terms a loan against the amount of money deposited with it at that time ... The bank has simply to agree that a person may take out a loan of, say, £5,000. The person taking out the loan can then spend £5,000 and hey presto! £5,000 of new number-money has been created ... Whoever takes out the loan will then make purchases and payments to other people, who will pay that new money into their bank accounts ... Total deposits in the banking system have therefore increased by £5,000. This is the boomerang effect of a bank loan by which a loan rapidly creates an equivalent amount of new bank deposits in the banking system. The new money will provide the banking system with the collateral for more lending. This is the bolstering effect of a bank loan. As the total money held by banks and building societies becomes swollen by loans returning as new deposits this provides them with the basis for further loans.
>
> (Rowbotham, 1998)

An important consequence of this system is that most of the money in circulation only exists because either businesses or individuals have gone into debt and are paying interest on their loans. Household mortgages play a key role here, accounting for around two-thirds of the money supply. While individual loans may be repaid, the debt in aggregate can never be repaid because this would remove virtually all the money from circulation. The health of the economy is therefore entirely dependent upon the continued willingness of businesses and consumers to take out loans for either investment or consumption (so-called 'consumer confidence'). Any reduction in borrowing therefore threatens to tip national economies into recession.

Individual loans need to be repaid with interest, but this removes money from circulation. Hence, the only way that individual borrowers can pay the interest on their loans, without at the same time reducing the amount of money in circulation, is if they, or other borrowers, borrow an equivalent amount more (Douthwaite, 2000). As a result, the amount of money in circulation needs to *rise* each year by an amount that is at least proportional to that being removed through interest payments. This means that the value of goods and services bought and sold must also rise, either through inflation or through higher consumption. In other words, the economy must grow.

For individual businesses, if there is no growth in any year, the investments made the previous year will have produced no return. Firms will then find themselves with lower profits and unused capacity, discouraging them from investing further, at least in those sectors in which the increased capacity has not been taken up. Less investment will mean fewer loans being taken out and thus less money entering into circulation to replace that being removed through interest payments. And less money in circulation will mean that there is less available for consumers to spend, which will very soon lead to economic slowdown, with the resulting bankruptcies and unemployment. By such processes, the monetary system creates a structural requirement for continued growth and increased consumption. As alternative economists, Richard Douthwaite and Emer Siochrú observe:

> This prospect of investment falling and creating widespread unemployment terrifies governments so much that they work very closely with their business sectors to ensure that their economies continue to grow almost regardless of any social or environmental damage the growth process may be causing. In other words, the need for growth to maintain short-term economic sustainability gets in the way of attending to more fundamental types of sustainability such as halting social decline or climate change.
>
> (Douthwaite and Siochrú, 2006)

The consequences of this dynamic are insufficiently recognised but crucial. The implication is that, under our current economic system, reducing consumption through sufficiency would have serious consequences for business, consumers and governments. As a result, the widespread adoption of the sufficiency strategy would face major political obstacles. Recognition of this is slowly growing, with proposals for monetary reform being developed by a number of groups including the Foundation for the Economics of Sustainability (Douthwaite and Siochrú, 2006) and the New Economics Foundation (Huber and Robertson, 2000). Monetary alternatives such as local currencies and time banks are also being pioneered at the local level, alongside the promotion of local markets and businesses (Ward and Lewis, 2002) and community-based initiatives covering a wide range of products and services (Seyfang, 2008). But unless the primary monetary system is reformed, the structural obstacles to the sufficiency strategy will remain.

Changing consumption to reduce carbon emissions

OECD countries have binding targets to reduce carbon emissions under the Kyoto Protocol and some countries, such as the UK, have adopted ambitious long-term goals for emissions reduction. It is hoped that these objectives will be achieved in part through lower energy consumption bought about by 'sustainable consumption'. Although energy efficiency figures prominently in official rhetoric, there is also mention of lifestyle change, to be achieved through either consuming differently or consuming less. But such changes in our energy habits are likely to require corresponding changes in our notions of comfort, convenience and cleanliness. While some limited progress is being made in these areas through the work of voluntary organisations and social movements, so far there has been little encouragement by governments. For instance the slow food movement is challenging our idea of convenience in food preparation and consumption – faster and quicker is not necessarily better (Honoré, 2004) – while organisations such as Global Action Plan and Car Clubs are challenging conventional ideas about what is comfortable, convenient and acceptable. The message is that recycling is not inconvenient, you can be comfortable not owning a car, and that changes to a green lifestyle are socially desirable.

But what contribution can such voluntary changes in consumption make towards meeting climate change targets? In the short term 'green consumerism' or adopting more efficient (or greener) products can reduce energy consumption, but the overall effect may soon be outweighed by continued economic growth. Research in Sweden by Eva Alfredsson (2004) shows that while adopting green consumption patterns can achieve a 10–20 per cent reduction in energy consumption in the short term, these reductions will soon be outpaced by rising levels of consumption, caused by modest growth in income (1–2 per cent per annum). Greater reductions in energy consumption are possible in the long term, but only if they are not outweighed by corresponding increases in income. The long-term potential of green lifestyles has been explored by Peter Harper (2007), who contrasts two differing households (with two adults and two children) in the year 2030 in terms of their ability to reduce carbon emissions. The Well-Off Techie environmentalists (the WOTs) are able to afford the most efficient technology (such as low-energy housing and 'hyper car') with minimal changes in lifestyle, allowing them to reduce their carbon emissions by a quarter. In contrast, the Low-Income Lifestyle environmentalists (the LILS) are able to reduce their emissions by over half, mainly through not using air travel,

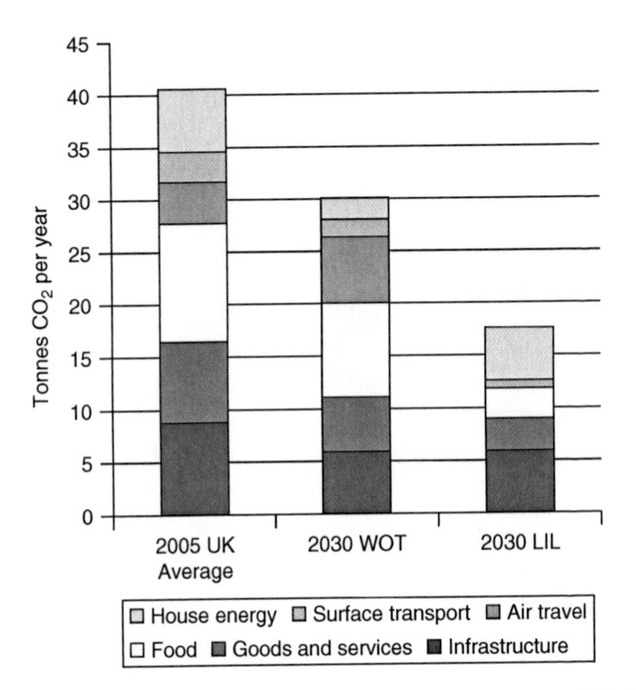

Figure 11.2 Emissions for hypothetical four-person household in 2030 (tonnes of carbon dioxide per annum)
Source: Harper (2007, reproduced with permission of Palgrave Macmillan).

only occasionally using cars, having a vegan diet and a much lower consumption of goods and services. The results are shown in Figure 11.2.

Harper's results suggest that, if average incomes increase to the level of the WOTs, energy efficiency alone will be insufficient to meet the UK government's target of a 60 per cent reduction in carbon emissions by 2050. In contrast, lifestyle changes by households at the income level of the LILs could potentially achieve it, but the implied lifestyle changes pose a major challenge to our accepted norms of comfort and convenience. For example, they include:

- indoor heating temperatures of 16–17°C;
- no car ownership;
- very little meat, fish or dairy consumption;
- no holidays abroad; and
- very limited expenditure on consumer goods.

This lifestyle of 'green austerity' is likely to be rejected by most people, as it is currently associated with extreme poverty or hippy eccentricity. However, the key point is that those practising and enjoying such a life generally do so not as individuals but as part of social groups with a common aim and shared vision (Simms and Smith, 2008). Their satisfaction or well-being comes more from their sense of belonging to a community and from living an ethical life, than from their ownership of material goods (Jackson, 2005). Thus the common thread to these lifestyle changes is the importance of community and social links. The relevant model is not the individual rational consumer, maximising satisfaction and meaning through exercising choice in the market. Instead, it is the individual as part of a community: of people finding meaning and satisfaction through leading 'a life without shame' – namely one that wins approval from families, friends and community (Jackson 2007: 373).

Conclusion

Attempts by individuals to lower their energy consumption are likely to have unintended consequences. There may be benefits for us personally and for our communities, but whether there will be benefits for our national economies or the planet is open to question. It is far too simplistic to argue that a *voluntary* reduction in our energy use would lead to lower *global* carbon emissions. That goal requires action by governments and coordinated international cooperation. In the long term, it will only be achieved through recognising the unsustainability of continued economic growth in rich countries and by addressing the structural causes of that growth, including in particular the debt-based monetary system. For individuals, it may be better to target consumption boycotts at specific products and to campaign for broader political reforms as has been done with great success historically in the campaigns to save wildlife and improve animal welfare (Adams, 2004).

Consumption patterns have been changed in the past through public campaigns, generally in areas where they were acceptable alternatives. Attempts to change demand patterns have generally been more successful than attempts to reduce demand. Moral statements about refusing to consume have always been powerful weapons in environmental and social campaigns, but victory has come about not by the bankruptcy of suppliers (due to decreased demand) but by changes in public opinion which have been sufficiently powerful to force reluctant governments to act on the supply side: to ban the transport of slaves, to prohibit the import of animal furs, to set welfare regulations for animals, or to set limits on carbon emissions.

As Blake Alcott aptly concludes (2008: 782):

> Efficiency and sufficiency strategies both offer relatively painless solutions to non-sustainability. The former is praised by 'negawatt' advocates not only as a free lunch, but one you are paid to eat. The latter appears to many of its addressees tolerable – switching off a few lights, riding a bike, or eating less meat, and here, too, a lunch you are paid to eat comes in the form of various health and happiness benefits. Supply-side or other impact-side strategies, on the other hand, are hard. They confront us with the neglected question of the carrying capacity of the planet.

This, the carrying capacity of the planet, is the crucial question for energy use and carbon emissions. Our desire for sufficiency, in a world of mass poverty, speaks to us of the futility of mass consumption and the desire for global justice. But practising sufficiency is not nearly enough of a response to the mainly technical and economic questions of how to reduce environmental damage. That damage cannot be abated by a minority of ethical consumers: it needs the actions of the majority, expressed through government action. Better, and more feasible, that we all consume slightly less than a few consume much less.

References

Adams, W. (2004) *Against Extinction: the Story of Conservation*, London: Earthscan.

Alcott, B. (2008) 'The sufficiency strategy: would rich-world frugality lower environmental impact?', *Ecological Economics*, 64(4): 770–86.

Alfredsson, E. (2004) '"Green" consumption – no solution for climate change', *Energy* 29(4): 513–24.

Caird, S. and R. Roy (2007) *Consumer Adoption and Use of Household Renewable Energy Products*, Design Innovation Group, Open University.

Douglas, M. (1966) *Purity and Danger: an Analysis of Concepts of Pollution and Taboo*, London: Routledge.

Douglas, M. and B. Isherwood (1979) *The World of Goods*, New York: Basic Books.

Douglas, M., D. Gasper, S. Ney and M. Thompson (1998) 'Human needs and wants', in S. Rayner and E. Malone (eds), *Human Choice and Climate Change*, Columbus, Ohio: Battelle Press, Volume 1, pp. 95–264.

Douthwaite, R. (2000) *The Ecology of Money*, Totnes, Devon: Green Books.

Douthwaite, R. and E. Siochrú (2006) *The Economic Challenge of Sustainability*, Dublin, Ireland: Foundation for the Economics of Sustainability, August. Available at www.feasta.org/money.htm

Duncan, G. (2008) 'Why oil rulers won't go green', *The Times*, 3 March, p. 39.

Fouquet, R. and P. Pearson (2006) 'Seven centuries of energy service: the price and use of light in the United Kingdom (1300–2000)', *Energy Journal*, 27(1): 139–76.
Ghazi, P. and J. Jones (2004) *Downshifting: the Bestselling Guide to Happier, Simpler Living*, London: Hodder & Stoughton.
Halme, M. (2005) *Sustainable Consumer Services: Business Solutions for Household Markets*, London: Earthscan.
Harper, P. (2007) 'Sustainable lifestyles of the future', in D. Elliott (ed.), *Sustainable Energy: Opportunities and Limitations*, Basingstoke: Palgrave Macmillan, pp. 236–60.
Hertwich, E. (2006) 'Accounting for sustainable consumption', in T. Jackson (ed.), *The Earthscan Reader in Sustainable Consumption*, London: Earthscan, pp. 88–108.
Honoré, C. (2004) *In Praise of Slow: How a Worldwide Movement is Challenging the Cult of Speed*, London: Orion Press.
Huber, J. and J. Robertson (2000) *Creating New Money: a Monetary Reform for the Information Age*, London: New Economics Foundation.
Jackson, T. (2005) 'Live better by consuming less? Is there a "double dividend" in sustainable consumption?', *Journal of Industrial Ecology*, 9(1–2): 19–36.
Jackson, T. (2007) 'Consuming paradise? Towards a social and cultural psychology of sustainable consumption', in T. Jackson (ed.), *The Earthscan Reader in Sustainable Consumption*, London: Earthscan, pp. 367–95.
Patterson, W. (1990) *The Energy Alternative*, London: Boxtree.
Petherick, T. (2007) *Sufficient: a Modern Guide to Sustainable Living*, London: Pavilion Books.
Princen, T. (2005) *The Logic of Sufficiency*, London: MIT Press.
Rejeski, D. (ed.) (2003) 'E-commerce, the Internet and the environment', Special issue of *Journal of Industrial Ecology*, 6(2).
Rowbotham, M. (1998) *The Grip of Death: a Study of Modern Money, Debt Slavery and Destructive Economics*, Charlbury, Oxfordshire: Jon Carpenter Publishing.
Roy, R. (2000) 'Sustainable product-service systems', *Futures*, 32(3–4): 289–99.
Seyfang, G. (2008) *Building Sustainable Communities*, Basingstoke: Palgrave Macmillan.
Shi, D. (1985) *The Simple Life: Plain Living and High Thinking in American Culture*, New York: Oxford University Press.
Shove, E. (2004) *Comfort, Cleanliness and Convenience: the Social Organization of Normality*, Oxford: Berg.
Simms, A. and J. Smith (eds) (2008) *Do Good Lives Have to Cost the Earth?* London: Constable.
Trainer, T. (1985) *Abandon Affluence*, London: Zed Books.
Ward, B. and J. Lewis (2002) *Plugging the Leaks: Making the Most of Every Pound that Enters your Local Economy*, London: New Economics Foundation. Available at www.pluggingtheleaks.org/handbook.pdf
Wilhite, H. (2005) 'Why energy needs anthropology', *Anthropology Today*, 21(3): 1–2.
Zavestoski, S. (2001) 'Environmental concern and anti-consumerism in the self-concept', in M. Cohen and J. Murphy (eds), *Exploring Sustainable Consumption: Environmental Policy and the Social Sciences*, Oxford: Pergammon, pp. 173–89.

12
Conclusion

Steve Sorrell and Horace Herring

Most informed people now acknowledge that carbon emissions will need to be rapidly and significantly reduced if dangerous climate change is to be avoided. The scale of the required reductions is daunting and is unlikely to be achievable without at the same time reducing energy demand, at least within the developed world (Pacala and Socolow, 2004). But despite decades of government efforts to improve energy efficiency, energy consumption continues to rise. For instance, in the residential sector of the European Union (the EU-25) over the period 1999–2004, total electricity consumption grew by 11 per cent, and gas by 14 per cent, despite numerous energy-efficiency polices at both the EU and national level (Bertoldi and Atanasiu, 2007). Since primary energy consumption is growing more slowly than GDP in OECD countries, it is frequently claimed that that energy consumption has been 'decoupled' from GDP (Ang and Liu, 2006).[1] But this is 'weak' rather than 'strong' decoupling since total energy consumption continues to increase.

A 'Factor 4' improvement in the energy efficiency of OECD countries seems perfectly feasible this century, even at current rates of improvement. However:

> on current trends we find no reason to believe that these improvements can counteract the tendency for energy consumption to grow. Even if energy consumed per unit of output were reduced by three-quarters or Factor Four, half a century of economic growth at 3% a year (slightly less than the global trend for the past quarter century) would more than quadruple output, leaving overall energy consumption unchanged.
>
> (RCEP, 2000: 6.139)

Historical improvements in energy efficiency may have reduced energy consumption below what it would have been without those improvements, but since the 'counterfactual' cannot be observed we can never be sure. In contrast, there is no dispute that such improvements have not reduced energy consumption in absolute terms. The reason, according to many commentators, is that we simply haven't tried hard enough: while many energy efficiency policies are in place, they are often small-scale, underfunded, unambitious and ineffectual. Indeed, past improvements in various measures of energy efficiency owe more to market-driven structural and technological change than to any action by governments (Schipper, 1987). If the cause lies in inadequate policies, the solution could be to redouble our efforts – to introduce more regulations, standards and financial support alongside innovative measures such as tradable white certificate schemes (e.g. Boardman, 2007). However, an equally plausible explanation for the failure to reduce energy consumption is that many of the potential energy savings have been 'taken back' by various rebound effects. In practice, both of these explanations are likely to be partially correct but our present state of knowledge does not allow their relative importance to be established. As a result, the significance of rebound effects remains in dispute.

This chapter attempts to summarise the main arguments in this book and to highlight some of the implications for energy and climate policy. It does so by advancing six increasingly controversial propositions. These are:

- rebound effects matter;
- rebound effects can be quantified;
- policy can mitigate rebound effects;
- Jevons' Paradox holds in important cases;
- efficiency needs to be combined with sufficiency;
- sustainability is incompatible with continued economic growth in rich countries.

These propositions are based in part upon the chapters in this book and in part upon broader research including in particular the literature review on rebound effects conducted by the UK Energy Research Centre (UKERC) (Sorrell, 2007). The views expressed in this chapter are those of the authors and are not necessarily shared by the other contributors to this book. The following sections discuss each of these propositions in turn.

Rebound effects matter

The continuing neglect of rebound effects by both policy-makers and energy researchers is remarkable. Such neglect would be justified if rebound effects could be shown to be negligible in the majority of cases. But the evidence presented in this book and elsewhere suggests that this is not the case.

It is true that the evidence for rebound effects is limited and inconclusive. While the evidence is better for direct effects than for indirect effects, it remains overwhelmingly focused on a limited number of consumer energy services within the OECD. Both direct and indirect rebound effects appear to vary widely between different technologies, sectors and income groups and in most cases cannot be quantified with much confidence. However, following a comprehensive review of this evidence (partly summarised in Chapters 2 and 7) the UKERC concludes that economy-wide rebound effects are generally *not* negligible and in some cases could exceed unity. Rebound effects therefore need to be taken seriously in policy appraisal.

For personal automotive transport, household heating and household cooling within the OECD, the direct rebound effect is likely to be less than 30 per cent and may be closer to 10 per cent for transport (see Chapter 2). Moreover, the direct rebound effect for these energy services is expected to decline in the future as demand saturates and income increases. This suggests that improvements in energy efficiency should achieve 70 per cent or more of the expected reduction in energy consumption for those services – although the existence of indirect effects means that the economy-wide reduction in energy consumption will be less. Also, these conclusions are subject to a number of important qualifications, including the dependence of direct rebound effects on household income, the neglect of 'marginal consumers' (i.e. those who previously did not have access to the relevant energy service) and the relatively limited time periods over which such effects have been studied. In addition, a number of authors have challenged the consensus in this area, including Frondel et al. in Chapter 3 who find much higher direct rebound effects for personal automotive transport.

The evidence for direct rebound effects for other consumer energy services is sparse, as is that for energy services in production. Theory suggests that direct rebound effects should be smaller for other consumer energy services since energy forms a relatively small proportion of total costs and therefore has little influence on purchase or operating decisions. But for producers, a much wider range of direct rebound effects can be expected for different types of energy-efficiency improvement, especially over the long term where such improvements contribute to output growth and industrial expansion. In principle, the own price elasticity of energy consumption for a particular energy service should provide an upper bound to the direct rebound effect for that service and since estimates of such elasticities are generally less than unity, this suggests that backfire is unlikely from direct rebound effects alone (Sorrell and Dimitropoulos, 2007a). However, energy consumption is usually measured for a collection of energy services and many elasticity estimates are confined to the short to medium term. Similarly, there are very few studies of direct rebound effects from energy-efficiency improvements in developing countries, although theoretical considerations suggest that direct rebound effects in these contexts could be larger than in the OECD. This is especially the case for the 1.6 billion households who currently lack access to electricity and the 2.5 billion who rely upon biomass for cooking. The adoption of energy-efficient technologies for heating, cooling and lighting should dramatically improve the welfare of such households, but to rely upon them as a means to reduce energy consumption would be misguided (Roy, 2000).

Quantitative estimates of indirect and economy-wide rebound effects are rare and the limited number of studies available provides an insufficient basis from which to draw any general conclusions (see Chapter 4). Perhaps the most important insight from this work is that the magnitude of such effects depends very much upon the sector in which the energy-efficiency improvement takes place and is sensitive to a number of variables. A handful of CGE modelling studies estimate economy-wide rebound effects of 40 per cent or more following energy-efficiency improvements by producers, with half of these studies predicting backfire (Allan et al., 2007). Such large numbers should give cause for concern, especially since they derive from 'pure' energy-efficiency improvements and therefore do not reflect any additional increase in energy consumption that may result from associated improvements in the productivity of other inputs. However, the small number of studies available, the diversity of approaches used and the variety of methodological weaknesses associated with CGE modelling all suggest the need for great caution when interpreting these results. Some of these weaknesses may be overcome by macroeconometric models of national economies and these have been used by Barker and Foxon (2006) to estimate an economy-wide rebound effect of 26 per cent from energy efficiency policies in the UK. However, there are a number of reasons why this figure could be an underestimate (Sorrell, 2007).

Overall, the evidence suggests that rebound effects vary widely but in many cases could be substantial. To overlook them altogether would be a serious mistake.

Rebound effects can be quantified

Given the potential importance of rebound effects, the existing evidence base is remarkably sparse. This is partly a consequence of the inherent difficulty of measuring or estimating such effects. For example, Roger Levett (Chapter 9) argues that the complexity of technological and social systems makes it impossible to predict the results of any intervention beyond the very short term. Similar scepticism is expressed by Princen:

> Efficiency narrowly targets a single variable, sometimes two, in a system otherwise made up of multiple variables interacting in complex, often unpredictable ways. If the system were *complicated* like a clock, the variables and their interactions would be predictable ... But when the system is *complex*, like an organism or a weather system or a stock market, perturbations are largely unpredictable, or of limited predictability. A push here produces a ripple there. A ripple produces a wave. Some waves dissipate, others join other waves to produce storms.
>
> (Princen, 2005: 110)

However, this is a general argument about the unpredictability of social systems, rather than a specific point about rebound effects. The same criticism could be made of attempts to estimate the impact of energy taxes, but this is now the subject of a large and insightful literature (Ekins and Barker, 2001). The same methodological tools used to investigate policy measures such as energy taxes may also be used to estimate direct, indirect and economy-wide rebound effects. While each of these tools has numerous weaknesses, they can nevertheless improve our understanding of the mechanisms that contribute to rebound effects as well as the relative importance of different variables and the conditions under which such effects are more or less likely to be large. Examples of various approaches are available in the literature and provide a good basis for further research.

In the case of direct rebound effects, quantitative estimates are contingent upon good data sets and would benefit from more robust methodologies that address the possible sources of bias identified in Chapter 2. This is particularly the case for quasi-experimental studies of household heating, where current evaluation practice is poor. It is also important to consider alternative measures of energy services in order to capture all of the relevant effects – for example, using tonne kilometres rather than vehicle kilometres as a measure of personal automotive transport to allow the effect of (efficiency-induced) increases in car sizes to be explored. Data permitting, econometric and quasi-experimental techniques could be used to estimate rebound effects for a wider range of consumer energy services – although the policy issue here is not so much short-term changes in utilisation patterns but long-term changes in the number and capacity of conversion devices. The use of time-budget data could also allow the 'rebound effect with respect to time' to be investigated, most notably in the area of transportation (see Chapter 8).

For indirect rebound effects, a combination of input-output and life-cycle analysis can be used to estimate the 'embodied energy' associated with various types of energy-efficiency improvement (e.g. the energy required to manufacture and install thermal insulation) (Sartori and Hestnes, 2007). Such estimates are available but need to be developed more systematically and the results incorporated into technology and policy appraisals. Embodied energy analysis can also be combined with econometric models of consumer behaviour to estimate the secondary effects from household energy-efficiency improvements – although this approach tends to be restricted to rather aggregate categories of consumer expenditure (Brännlund et al., 2007).

Economy-wide rebound effects and the consequences of energy-efficiency improvements by producers may be best explored through energy-economic modelling. Given the widespread use of CGE models in energy research, the lack of application to rebound effects is surprising. In their review of this topic, Allan et al. (2007) identified only eight studies, several of which were methodologically weak (see Chapter 4). There is therefore considerable scope for expanding this evidence base, both in terms of the countries and sectors studied and also the types of energy-efficiency improvement that are modelled. This would also allow the results of different studies to be more meaningfully compared. Given the weak empirical basis of CGE models, it is essential that such studies include systematic and informed sensitivity analysis. More robust estimates of economy-wide effects could also be obtained from macro-econometric models (Barker and Foxon, 2006).

In summary, while precision is unobtainable, attempts to quantify rebound effects should be able to indicate approximate orders of magnitude as well as providing policy-relevant insights into the mechanisms involved. They should therefore be pursued.

Policy can mitigate rebound effects

Energy efficiency may be encouraged through policies that raise energy prices, such as carbon taxes, or through non-price policies such as building regulations. Appropriately designed, such policies should continue to play an important role in energy and climate policy. However, the neglect of rebound effects has meant that many official and independent appraisals have overstated the potential contribution of non-price policies to reducing energy consumption and carbon emissions. Taking rebound effects into account will reduce the apparent effectiveness of such policies.

At the same time, many energy efficiency opportunities appear to be highly cost-effective for the individuals and organisations involved (IPCC, 2007a). If non-price policies can cost-effectively overcome the various market and organisational failures that prevent such opportunities from being taken up, they should increase the real income and welfare of consumers and improve productivity in both the public and private sector. If economy-wide rebound effects are less than unity, they may also be cost-effective from a climate change perspective – for example, on the basis of the marginal abatement cost of carbon emissions. They may not, however, reduce energy consumption and carbon emissions by as much as previously assumed.

While rebound effects cannot be quantified with much confidence, there should be scope for including estimated effects within policy appraisals and for building 'headroom' into policy targets to allow for such effects. There may also be scope for using these estimates to target policies more effectively – for example, by prioritising sectors and technologies where rebound effects are expected to be smaller. In contrast, where rebound effects are expected to be large, there may be a greater need for policies that increase energy prices.

A particularly important finding from the UKERC review is that technologies that reduce capital and labour costs as well as energy costs may be associated with the largest rebound effects. This includes the 'win-win' opportunities that are regularly highlighted by energy efficiency advocates. For example, Lovins and Lovins (1997) used case studies to argue that better visual, acoustic and thermal comfort in well-designed, energy-efficient buildings can improve labour productivity by as much as 16 per cent. Since labour costs in commercial buildings are typically twenty-five times greater than energy costs, the resulting cost savings can dwarf those from reduced energy consumption. But if the total cost savings are twenty-five times greater, the indirect rebound effect may be twenty-five times greater as well. While identifying such situations is far from straightforward, the implications of encouraging such opportunities should ideally be taken into account. It may make more sense to focus non-price policy on 'dedicated' energy-efficient technologies, such as thermal insulation, that do not have these wider benefits.

The primary source of rebound effects is the reduced cost of energy services. In principle, policies such as carbon taxes should reduce direct and indirect rebound effects by ensuring that the cost of energy services remains relatively constant while energy efficiency improves. But carbon pricing will need to increase over time at a rate sufficient to accommodate both income growth and rebound effects, simply to prevent carbon emissions from increasing. It will need to increase more rapidly if emissions are to be reduced (Birol and Keppler, 2000).

Perhaps the most attractive approach to mitigating rebound effects is the use of cap and trade schemes for carbon emissions. The impact of these may helpfully be illustrated with the use of the so-called IPAT equation (Chertow, 2001). For example, economy-wide carbon emissions (I) may be expressed as the product of population (P), GDP per capita ($A = Y/P$) and carbon emissions per unit of GDP ($T = C/Y$): $I = PAT$. Non-price energy-efficiency policies seek to reduce carbon intensity (T) and thereby carbon emissions (I). But there may be feedback effects between the right-hand side variables in the form of various rebound effects (Alcott, 2008). In particular, energy efficiency may encourage economic growth (A), which in turn will increase the total demand for energy and hence carbon emissions (I). Improvements in energy efficiency could therefore have unintended consequences that undermine the policy objective. In contrast, by placing a quantitative limit upon carbon emissions, a cap and trade scheme focuses directly on the relevant environmental impact (I) rather than on one of the factors contributing to that impact. Provided the scheme is effectively enforced, it should provide a guarantee that the desired environmental outcome is achieved (although there may be unintended consequences for sources of emissions that are not covered by the scheme). A cap and trade scheme therefore provides some insurance against rebound effects. The downside, of course, is that a cap on carbon emissions (I) could have unintended consequences for the right-hand side variables, notably per capita GDP (A). The importance of this is greatly disputed but will depend in part upon how the scheme is implemented – including in particular the choice between free allocation of allowances and revenue neutral auctioning (Bovenberg, 1999).

Policies such as carbon taxes and cap and trade schemes may be insufficient on their own, since they will not overcome the numerous barriers to the innovation and diffusion of low-carbon technologies and could in some circumstances have adverse impacts on income distribution and competitiveness (Sorrell and Sijm, 2003). Similarly, non-price policies to address market barriers may also be insufficient, since rebound effects could offset some or all of the energy savings. Effective climate policy will therefore need to combine both.

Jevons' Paradox holds in important cases

The ghost of Jevons haunts modern-day climate policy. If Jevons' Paradox, or its modern variant the Khazzoom-Brookes (K-B) postulate, proves to be universally correct, energy efficiency policy will be doomed to failure – unless it is combined with a rejection of continued economic growth (see below). If, on the other hand, it only applies in certain situations, a compromise between environmental protection and continued economic growth appears feasible. The stakes could not be higher.

The K-B postulate does not lend itself to straightforward empirical test and many of the quantitative estimates of economy-wide rebound effects summarised above appear to contradict it. Suggestive and indirect support for the K-B postulate may be obtained from economic theory (see Chapter 5), mathematical modelling, econometric analysis and economic history, but these appear insufficient to verify that the 'K-B hypothesis' holds in all cases (see Chapter 7). However, these arguments and evidence do pose an important challenge to conventional wisdom, notably with respect to the contribution of energy to economic growth. For example, it is normally assumed that there is considerable scope for substituting capital and other inputs for energy and that technical change has improved the energy efficiency of individual sectors and thereby contributed to the observed decoupling of energy consumption from economic growth. But the evidence presented in Sorrell and Dimitropoulos (2007b) and summarised in Chapter 7 suggests that there is limited scope for substituting other inputs for energy and that much technical change has acted to increase energy intensity. Also, once different fuels are weighted by their relative economic productivity, there is less evidence that the growth in economic output has been decoupled from the growth in energy consumption. Overall, this evidence points to economy-wide rebound effects being relatively large and to energy playing a more important role in economic growth than is conventionally assumed.

A standard argument against the K-B postulate is that rebound effects must be small because the share of energy in total costs is small. But the possibility of large rebound effects becomes more plausible if it is accepted that energy-efficiency improvements are frequently associated with proportionately greater improvements in total factor productivity. This is a core theme of Len Brookes' work and also receives support from the work of Beaudreau (1995), Kummel et al. (2000) and Ayres and Warr (2005), who each estimate the productivity of energy inputs to be much greater than the share of energy in total costs.[2] Ayres and Warr go so far as to claim that it is energy efficiency, rather than technical innovation more generally, that is the primary driver of economic growth (Chapter 6). However, Sorrell and Dimitropoulos (2007b) highlight some weaknesses with this argument and emphasise that such productivity benefits need not be associated with *all* types of energy-efficiency improvement. Instead, the link between energy efficiency and total factor productivity is likely to be contingent upon particular technologies and circumstances.

As argued in Chapter 7, the debate over the K-B postulate would benefit from more careful distinctions between different types of energy-efficiency improvement. For example, the K-B postulate seems more likely to hold for energy-efficiency improvements associated with so-called general-purpose technologies (GPTs), particularly when these are used by producers and when the improvements occur at an early stage of development and diffusion. Steam engines provide a paradigmatic illustration of a GPT in the nineteenth century, while electric motors provide a comparable illustration for the early twentieth century. The opportunities offered by these technologies have such long-term and significant effects on innovation, productivity and economic growth that economy-wide energy consumption is increased. In contrast, the K-B postulate seems less likely to hold for dedicated energy-efficiency technologies such as thermal insulation, particularly when these are used by consumers. These technologies have smaller effects on productivity and economic growth, with the result that economy-wide energy consumption may be reduced – at least, compared to a scenario in which such technologies are not introduced. The relative importance of these different categories remains to be established, but it seems likely that the K-B postulate holds for at least some important types of energy-efficiency improvement. Whatever the other benefits of such technologies, encouraging them as a means to mitigate climate change could be counterproductive unless combined with effective carbon/energy pricing.

Efficiency needs to be combined with sufficiency

The achievement of much higher levels of energy efficiency seems essential if carbon emissions are to be radically reduced. But given the potential importance of rebound effects, this may not be sufficient to meet ambitious carbon targets.

Government promotion of energy efficiency dates back to the 1970s, when there was much concern about the 'limits to growth'. Studies such as Meadows et al. (1972) highlighted the risk of future economic collapse due to ecological and resource constraints, but underestimated the potential of substitution and technical change to mitigate environmental impacts – at least in the short term (Cole et al., 1973). The proposed solution of a 'steady-state economy' (Daly, 1973) found little support and failed to be translated into concrete political strategies. Environmental problems accumulated nevertheless and gained increasing political attention. In 1987, the Brundtland Commission introduced the notion of sustainable development and proposed that economic growth could be reconciled with environmental protection, largely through the medium of resource efficiency (WCED, 1987). This message proved more acceptable to established interests and the inherent ambiguity of sustainable development facilitated the widespread acceptance of the idea – although without noticeably slowing the rate of either resource depletion or environmental destruction. Sustainability is now the stated goal of many governments, including the UK which has adopted sustainable consumption and production (SCP) as one of the guiding principles of its sustainable development strategy (DEFRA, 2005). The UK government defines SCP as 'doing more with less', which implies that it is synonymous with resource efficiency.[3]

The preferred strategy to achieve sustainable consumption is *consuming more efficiently*, which implies reducing the environmental impacts associated with each good or service. But the extent to which this is successful will depend upon the size of any associated rebound effects. If these are significant, resource and energy use may not be reduced as much as expected and in some circumstances could increase. Despite this, most OECD countries pay little attention to such possibilities and offer few options for mitigating the undesirable consequences.

A related and complementary strategy is *consuming differently*, which implies shifting towards goods or services with a lower environmental impact. This could involve purchasing 'greener products', increasing expenditure on 'services' rather than manufactured goods, or entering into arrangements such as energy service contracting and car sharing schemes. These strategies are frequently cited as having environmental benefits, although the empirical evidence to support such claims is often lacking. For example, a shift towards services and away from products may increase resource use, particularly if it involves higher standards of services, the extensive use of transport, or resource-intensive infrastructure such as telecommunications networks (e.g. home deliveries of internet shopping or take-away meals). In a recent review, Heiskanen and Jalas (2003) concluded that the environmental benefits of such arrangements are modest at best while Suh (2006) estimated that a shift to a service-oriented economy could *increase* carbon emissions (although reduce the carbon intensity of GDP). In Chapter 8, Jalas shows how consuming differently could also be associated with rebound effects, in this case in relation to the outsourcing of household services and the corresponding adjustments in households' use of time.

Given the potential limitations of consuming efficiently and consuming differently, it seems logical to examine the potential of a third option – simply *consuming less* (Schor, 1999; Sanne, 2000; Princen, 2005). This strategy has been advocated by number of authors, including Nørgård in Chapter 10. The key idea here is *sufficiency*, defined by Princen (2005) as a social organising principle that builds upon established notions such as restraint and moderation to provide rules for guiding collective behaviour. The primary objective is to respect ecological constraints, although most authors also emphasise the social and psychological benefits to be obtained from consuming less.

While Princen (2005) cites examples of sufficiency being put into practice by communities and organisations, most authors focus on the implications for individuals. They argue that 'downshifting' can both lower environmental impacts and improve quality of life, notably by reducing stress and allowing more leisure time (Chapter 10). This argument is supported by the results of several recent studies, which show that reported levels of happiness are not increasing in line with income in developed countries (Easterlin, 2001; Diener and Biswas-Diener, 2002; Blanchflower and Oswald, 2004). As Binswanger (2006: 377) observes 'the economies of developed countries turn into big treadmills where people tried to walk faster and faster in order to reach a higher level of happiness but in fact never get beyond their current position. On average, happiness always stays the same, no matter how fast people are walking on the treadmills.' The treadmill is a metaphor for the pursuit of more happiness by striving for more income, which leads to more income but not necessarily more happiness. It is possible that an ethic of sufficiency could provide a means of escaping from such treadmills while at the same time contributing to environmental sustainability.

According to Alcott (2008: 771), sufficiency implies both environmental motivation and purchasing power: 'those who are to alter their behaviour towards less consumption must be *able to* consume. Their purchasing power either remains unused or is itself reduced through working and earning less.' Hence, the concept appears primarily applicable to the wealthy in developed countries and is of little relevance for those suffering from absolute or relative poverty (although the appropriate threshold level of income is unclear). Given increasing income inequality in most OECD countries and spiralling levels of personal debt, the proportion of people able to exercise such a choice could be falling.

Adopting sufficiency as a guiding principle would involve a considerable change in lifestyles. While most people would acknowledge that 'quality of life' is not solely about material consumption, numerous social, psychological, economic and cultural obstacles can make it difficult to reduce current levels of consumption. For example, many people are partially 'locked in' to current consumption patterns owing to factors such as land-use patterns and physical infrastructures (which constrain choice in areas such as travel), the rapid obsolescence of consumer goods, the difficulty of reducing the number of hours of work, the never-ending search for status through the acquisition of symbolic 'positional goods', and the rapid adaptation of aspirations to higher income levels, thereby reducing the happiness associated with that income (Hirsch, 1995; Sanne, 2002; Binswanger, 2006). In this context, the practice of sufficiency requires a minimum level of financial security, deeply held values and considerable determination. If adopted successfully by enough individuals, it could demonstrate a viable and attractive alternative to consumerism and begin to shift social attitudes in a number of areas. But since 'luxury' consumption fulfils so many deep psychological needs, the widespread adoption of sufficiency appears unlikely to develop through voluntary action alone.

If the balance of factors encouraging or discouraging sufficiency were to change in favour of the former, it could become complementary to efficiency as a means of both improving quality of life and adapting to tightening ecological constraints. But the development of a 'sufficiency ethic' is likely to require *collectively agreed* constraints imposed by governments in some form. The personal carbon allowance scheme currently being discussed in the UK is a possible example (Hillman and Fawcett, 2004). This means that the most important agent of change is likely to be individuals acting as citizens in the political process rather than as 'downshifting' consumers. Moreover, as Alcott (2008) has pointed out, sufficiency is not immune from rebound effects, even when pursued at a national level. A successful 'sufficiency strategy' will reduce the demand for energy and resources, thereby lowering prices and encouraging increased demand by others which will partly offset the energy and resource savings (see Chapter 11). There are millions of (marginal) consumers worldwide who would happily increase their consumption if the prices of commodities fell. While this 'sufficiency rebound' is unlikely to exceed unity, it will nevertheless reduce the environmental benefits of any 'sufficiency strategy'. Again, to effectively address problems such as climate change, collective agreement is required, in this case at the international level.

Such international agreements should encourage the development and deployment of low and zero-carbon energy sources (e.g. renewables, nuclear or coal with carbon sequestration). These may initially be more expensive than fossil fuels, but costs should fall in the medium to long term due to scale and learning economies. If the cost savings from improved energy efficiency can be recycled to finance such investments, then this may be a relatively painless way of reducing the growth in material consumption and shifting to a low-carbon economy.

Of course, all energy sources have some adverse environmental impacts. For example, wind farms attract vigorous opposition owing to their visual impacts, while biomass plantations can displace both food production and 'wild' nature. Also, all energy sources are both resource-limited and require the consumption of energy, such as in the extraction, processing and transport of liquid fuels, in order to provide the energy in an economically useful form. Since this so-called energy return on investment (EROI) tends to be lower for renewable sources than for conventional oil and gas, it is possible that less 'net energy' will be available in the future for driving economic growth. Thus, while combining energy efficiency with low carbon sources may offer a breathing space for a few decades, it does not address the longer-term issue of limiting the growth in energy consumption before we reach non-energy eco-limits, such as lack of water, land and resources (Elliott, 2003).

Sustainability is incompatible with continued economic growth in rich countries

In much of this book, rebound effects have been judged negatively as they interfere with the objective of reducing energy consumption and hence carbon emissions. But from a conventional perspective, rebound effects are positive, since they encourage increased consumption and economic growth. The rebound effect is simply the natural outcome of efficiency, or productivity improvements in market economies, where cost savings are channelled into increased output and consumption under the assumption that this will improve human welfare. As a result, attempts to restrain rebound effects through, for example, carbon pricing could potentially damage economic growth (see Chapter 6).

The extent to which this will occur is very uncertain and depends upon the scope for substituting away from energy-intensive activities and carbon-intensive fuels. Nevertheless, at the deepest level rebound effects highlight a potential conflict between reducing energy use in absolute terms while at the same time continuing to grow the economy. Recognising the importance of rebound effects could therefore reopen the debate about various limits to growth.

The critique of economic growth is multifaceted and impossible to summarise here (see for example Douthwaite, 1999). But the key point is that the goal of economic development should not be to maximise GDP but to improve human well-being and quality of life. Material consumption is merely a means to that end and GDP is merely a measure of that means. It is likely that, beyond a certain level increased material consumption will reduce well-being, since the (typically unmeasured) social and environmental costs will exceed the benefits. Human well-being is not determined solely by the consumption of goods and services but also by 'human capital' (e.g. health, knowledge), 'social capital' (e.g. family, friends and social networks) and 'natural capital' (i.e. ecosystems and the services they provide) – none of which are necessarily correlated with GDP (Ekins, 1992). Attempts to value these contributions through the use of alternative measures of economic progress typically find that 'well-being' is not improving or even declining in rich countries, despite increases in GDP (Hanley et al., 1999; Nourry, 2008). Hence, while growth in per capita income is likely to improve well-being in developing countries, the same may not be true for developed countries. Table 12.1, which is taken from Costanza (2008), compares the emerging 'green' perspective on economic development with the current development model. While many elements of this perspective are now influencing mainstream policy (for example, the pricing of environment externalities and the development of alternative indicators of economic welfare), conventional assumptions still dominate, most notably at the macroeconomic level where maximising GDP remains the dominant goal.

Table 12.1 Different models of economic development

	Conventional	*'Green'*
Primary policy goal	Economic growth as measured by GDP. Growth should allow the solution of other problems	Development in the sense of improved quality of life. Growth has negative side-effects
Primary measure of progress	Gross Domestic Product (GDP)	Index of Sustainable Economic Welfare (ISEW) or something similar
Scale/carrying capacity	Not an issue because it is assumed that market could overcome resource limits via substitution and technical change	Primary concern since there is limited scope for substituting natural for man-made capital
Income distribution	Secondary concern. A 'trickle down' policy (a rising tide lifts all boats)	Primary concern. Directly affects quality of life and is often made worse by economic growth
Economic efficiency/ allocation	Primary concern, but generally including only marketed goods and services	Primary concern, but including both market and non-market goods and services. Natural and social capital must be valued
Role of 'sufficiency'	Not recognised. More is always better	Congruent with overall aims. More is not always better

Source: Adapted from Costanza (2008).

Over the long term, continued economic growth can only be reconciled with environmental sustainability if implausibly large improvements in energy and resource efficiency can be achieved.[4] Take, for example, the goal of keeping the long-term equilibrium rise in global temperature to below two degrees centigrade. According the IPCC (2007b), this will require a reduction in global carbon dioxide emissions (compared to 2000) of between 50 and 85 per cent by 2050. But the UN forecasts that the world's population will be close to nine billion by 2050. If we assume a 2 per cent annual increase in global per capita GDP (close to the average in recent years), this means that carbon emissions per unit of GDP will need to fall by at least 87 per cent by 2050 if carbon emissions are to be halved. If an 85 per cent reduction in carbon emissions is required, the corresponding figure is 96.3 per cent. If global per capita GDP grows at 3 per cent per annum, the corresponding figure is 97.7 per cent. Since we would expect rich countries to contribute a proportionally greater share of the emission reductions, they would need to make a correspondingly greater reduction in their carbon intensities. The efficiency strategy looks even less plausible if we extend the time horizon to 2100. A further half-century of economic growth at 2 per cent per annum would leave per capita GDP in 2100 more than seven times higher than in 2000. But to avoid dangerous climate change, further substantial cuts in global carbon emissions are likely to be required after 2050. Even if emissions in 2100 were simply to be stabilised at 15 per cent of 2000 levels (and assuming the global population remained at nine billion), global carbon emissions per unit of GDP in 2100 would need to be no more than 1 per cent of what they were in 2000.

Reductions of this scale may be possible and a number of economic models suggest that they can be achieved, at least up to 2050 (IPCC, 2007a). But there are reasons to be sceptical. For example, while global carbon emissions per unit of GDP have fallen by around 30 per cent since 2005, the rate of improvement has been declining and has recently reversed (Raupach et al., 2007). The historical improvements in this ratio owe much to the shift towards higher quality fuels such as natural gas (Kaufmann, 1992), but accelerating resource depletion could lead to an increased reliance on carbon-intensive fuels such as coal and non-conventional oil (Brandt and Farrell, 2007). The above figures could be made more plausible if the emissions target was relaxed, but the scientific consensus is that this is likely to put the two degrees target out of reach and thereby trigger serious consequences, such as the irreversible melting of the Greenland ice sheet (IPCC, 2007b; Lenton et al., 2007). Alternatively, the figures could be made more plausible if the assumption of a 2 per cent annual growth in per capita GDP was abandoned – or at least, if lower or zero levels of per-capita GDP growth were considered acceptable for developed countries (Ekins, 1993). But this runs counter to the current objectives of all developed country governments.

Hence, in an increasingly 'full' world, the sustainability of conventional economic growth in the rich countries deserves to be questioned. Continued growth may not be compatible with either environmental constraints or improvements in the quality of life. But while this view is gaining increasing acceptance, so far it has had practically no influence on public policy. There are a number of reasons for this, but the main one is that modern economies are structurally dependent upon continued economic growth (as measured by GDP) which makes any questioning of that growth unacceptable:

> During the last two centuries we have known nothing but exponential growth and in parallel we have evolved what amounts to an

> exponential growth culture, a culture so heavily dependent upon the continuance of exponential growth for its stability that it is incapable of reckoning with problems of non-growth.
>
> (Hubbert, 1976)

The reasons for this 'growth addiction' are varied but could lie in particular with the nature of our financial system. As noted in Chapter 11, the bulk of the money supply is created by private banks in the form of interest-bearing loans. These loans have to be repaid, which requires additional money which in turn requires more loans and hence more economic growth (Douthwaite, 2006). Since any slowing of the rate of growth has serious consequences for businesses (e.g. losses, bankruptcies), individuals (e.g. defaults on loans, unemployment) and politicians (e.g. loss of office) all sectors of society have strong incentives to maximise economic growth. Set against this, any concerns about long-term environmental sustainability and quality of life can easily be overridden. Hence, the challenge is not simply to demonstrate the unsustainability of the present model of economic development and the benefits of alternative models (Table 12.1) but to propose ways in which the dependence of modern economies upon continued economic growth can be broken – for example, through monetary reform. Time may be short, but that work has hardly begun.

Conclusion

To date rebound effects have been neglected by both researchers and policy-makers and where they have been considered, the focus has tended to be upon narrow technical issues such as the own-price elasticity of various energy services. But this is only one aspect of the problem. On another level, rebound effects raise fundamental questions about the sustainability of economic growth and the contribution of energy to that growth. Probably because this poses such a challenge to established interests and conventional wisdom, the deeper implications of rebound effects have been largely ignored.

This book has tried to illustrate all aspects of the rebound effect – from the microeconomic issues (Chapters 2 and 3), to the broader macroeconomic impacts (Chapters 4, 5, 6 and 7) and the deeper social and cultural implications (Chapters 8, 9, 10 and 11). This overview has led us to advance six propositions, namely: rebound effects matter; rebound effects can be quantified; policy can mitigate rebound effects; Jevons' Paradox holds in important cases; efficiency needs to be combined with sufficiency; and sustainability is incompatible with continued economic growth in rich countries. The first of these is the least controversial, least normative and best supported by the available evidence, while the opposite is the case for the last. But all of them deserve serious consideration.

It is frequently assumed that to highlight the potential importance of rebound effects is to question the importance of encouraging energy efficiency. Brookes (1990), for example, uses rebound effects to criticise the rationale of government energy efficiency policy. This is not our position. As we noted in Chapter 1, we cannot conceive of a sustainable economy that does not involve much higher levels of energy efficiency than exist today. We also consider that policies to encourage energy efficiency have an important role to play alongside carbon/energy pricing. What we find more difficult to accept is that deep cuts in carbon emissions can be achieved while indefinitely maintaining high rates of economic growth in the developed world. While we do not have a blueprint for achieving a sustainable economy, we consider that the existence of non-trivial rebound effects makes it all the more urgent that such fundamental questions be asked.

The rebound effect is therefore of critical importance for sustainability. If we continue to channel the benefits of energy efficiency into more growth and greater consumption, we run the risk of catastrophic environmental impacts. If instead, we channel these benefits into lower carbon energy supply and improved quality of life, it is just possible that catastrophe can be avoided.

Notes

1. For example, while the aggregate energy intensity of the UK economy improved by a factor of two between 1970 and 2006, total energy consumption increased by 11 per cent as a result of GDP more than doubling.
2. Beaudreau (1995) estimates the output elasticity for electric power in US manufacturing to be 0.53, which implies that a 1 per cent increase in electric power consumption results in a 0.53 per cent increase in manufacturing value added. This is much larger than conventional estimates: for example, Berndt and Wood (1975) estimate aggregate energy to have an output elasticity of only 0.06 in US manufacturing.
3. It also proposes measuring SCP through indicators of resource use and environmental impacts that are defined *relative* to GDP, without specifying whether weak or strong decoupling is required. However, the strategy does include some long-term environmental targets, most notably on CO_2 emissions.
4. Widespread neglect of this fact has led Bartlett (2004) to comment that 'The greatest shortcoming of the human race is our inability to understand the exponential function.'

References

Alcott, B. (2008) 'The sufficiency strategy: would rich-world frugality lower environmental impact?', *Ecological Economics*, 64(4): 770–86.

Allan, G., M. Gilmartin, P. G. McGregor, K. Swales and K. Turner (2007) *UKERC Review of Evidence for the Rebound Effect: Technical Report 4: Energy-economic Modelling Studies*, UK Energy Research Centre, London.

Ang, B. W. and N. Liu (2006) 'A cross-country analysis of aggregate energy and carbon intensities', *Energy Policy*, 34(15): 2398–2404.

Ayres, R. U. and B. Warr (2005) 'Accounting for growth: the role of physical work', *Structural Change and Economic Dynamics*, 16(2): 181–209.

Barker, T. and T. Foxon (2006) *The Macroeconomic Rebound Effect and the UK Economy*, Report to the Department of the Environment, Food and Rural Affairs, 4CMR, Cambridge.

Bartlett, A. (2004) *The Essential Exponential: For the Future of Our Planet*, Lincoln, Nebraska: University of Nebraska Press.

Beaudreau, B. C. (1995) 'The impact of electric power on productivity: a study of US manufacturing 1950–1984', *Energy Economics*, 17(3): 231–6.

Berndt, E. R. and D. O. Wood (1975) 'Technology, prices, and the derived demand for energy', *Review of Economics and Statistics*, 57(3): 259–68.

Bertoldi, P. and B. Atanasiu (2007) *Electricity Consumption and Efficiency Trends in the Enlarged European Union*, Status Report 2006 (EUR 22753 EN), Brussels: European Commission.

Binswanger, M. (2006) 'Why does income growth fail to make us happier? Searching for the treadmills behind the paradox of happiness', *Journal of Socio-Economics*, 35(2): 366–81.

Birol, F. and J. H. Keppler (2000) 'Prices, technology development and the rebound effect', *Energy Policy*, 28(6–7): 457–69.

Blanchflower, D. G. and A. J. Oswald (2004) 'Well-being over time in Britain and the USA', *Journal of Public Economics*, 88: 1359–86.

Boardman, B. (2007) *Home Truths: a Low-carbon Strategy to Reduce UK Housing Emissions by 80% by 2050*, Environmental Change Institute, University of Oxford, Oxford.

Bovenberg, A. L. (1999) 'Green tax reforms and the double dividend: updated reader's guide', *International Tax and Public Finance*, 6: 421–43.

Brandt, A. R. and A. E. Farrell (2007) 'Scraping the bottom of the barrel: greenhouse gas emission consequences of a transition to low-quality and synthetic petroleum resources', *Climatic Change*, in press.

Brännlund, R., T. Ghalwash and J. Nordstrom (2007) 'Increased energy efficiency and the rebound effect: effects on consumption and emissions', *Energy Economics*, 29(1): 1–17.

Brookes, L. G. (1990) 'The greenhouse effect: the fallacies in the energy efficiency solution', *Energy Policy*, 18(2): 199–201.

Chertow, M. R. (2001) 'The IPAT equation and its variants: changing views of technology and environmental impacts', *Journal of Industrial Ecology*, 4(4): 13–29.

Cole, H. S. D., M. Johada, C. Freeman and K. L. R. Pavitt (1973) *Models of Doom: a Critique of the Limits to Growth*, New York: Universe Books.

Costanza, R. (2008) 'Stewardship for a "full" world', *Current History*, 107: 30–5.

Daly, H. E. (1973) *Toward a Steady-state Economy*, San Francisco: W. H. Freeman.

DEFRA (2005) *Securing the Future: the UK Government Sustainable Development Strategy*, London: H.M. Government.
Diener, E. and R. Biswas-Diener (2002) 'Will money increase subjective well-being?', *Social Indicators Research*, 57: 119–69.
Douthwaite, R. (1999) *The Growth Illusion: How Economic Growth Has Enriched the Few, Impoverished the Many and Endangered the Planet*, Gabriola Island, BC, Canada: New Society Publishers.
Douthwaite, R. (2006) *The Ecology of Money*, Dublin, Ireland: Foundation for the Economics of Sustainability.
Easterlin, R. A. (2001) 'Income and happiness: towards a unified theory', *Economic Journal*, 111: 465–84.
Ekins, P. (1992) 'A four-capital model of wealth creation', in P. Ekins and M. Max-Neef (eds), *Real-Life Economics: Understanding Wealth Creation*, London: Routledge.
Ekins, P. (1993) 'Limits to growth and sustainable development: grappling with ecological realities', *Ecological Economics*, 8(3): 269–88.
Ekins, P. and T. Barker (2001) 'Carbon taxes and carbon emissions trading', *Journal of Economic Surveys*, 15(3): 325–76.
Elliott, D. (2003) 'A sustainable future: the limits to renewables', in R. Douthwaite (ed.), *Before the Wells Run Dry: Ireland's Transition to Renewable Energy*, Totnes: Green Books, pp. 72–80.
Hanley, N., I. Moffatt, R. Faichney and M. Wilson (1999) 'Measuring sustainability: a time series of alternative indicators for Scotland', *Ecological Economics*, 28(1): 55–73.
Heiskanen, E. and M. Jalas (2003) 'Can services lead to radical eco-efficiency improvements? A review of the debate and evidence', *Corporate Social Responsibility and Environmental Management*, 10: 186–98.
Hillman, M. and T. Fawcett (2004) *How We Can Save the Planet*, London: Penguin.
Hirsch, F. (1995) *Social Limits to Growth*, London: Routledge.
Hubbert, M. K. (1976) 'Exponential growth as a transient phenomenon in human history. The fragile earth: towards strategies for survival', paper presented to the World Wildlife Fund, Fourth International Congress, San Francisco, See: http://www.hubbertpeak.com/hubbert/wwf1976/print.htm
IPCC (2007a) *Climate Change 2007: Mitigation of Climate Change*, Working Group III contribution to the IPCC Fourth Assessment Report, Intergovernmental Panel on Climate Change.
IPCC (2007b) *Climate Change 2007: Synthesis Report*, Fourth Assessment Report, Intergovernmental Panel on Climate Change.
Kaufmann, R. K. (1992) 'A biophysical analysis of the energy/real GDP ratio: implications for substitution and technical change', *Ecological Economics*, 6(1): 35–56.
Kummel, R., D. Lindenberger and W. Eichhorn (2000) 'The productive power of energy and economic evolution', *Indian Journal of Applied Economics*, 8: 231–62.
Lenton, T., H. Hermann, K. Elmar, J. W. Hall, W. Lucht, S. Rahmstorf and H. J. Schellnhuber (2007) 'Tipping elements in the Earth's climate system', *Proceedings of the National Academy of Sciences*, 105(6): 1786–93.
Lovins, A. B. and L. H. Lovins (1997) *Climate: Making Sense and Making Money*, Old Snowmass, Colorado: Rocky Mountain Institute.
Meadows, D. H., D. L. Meadows, J. Randers and W. W. Behrens III (1972) *The Limits to Growth*, New York: Universe Books.

Nourry, M. (2008) 'Measuring sustainable development: some empirical evidence for France from eight alternative indicators', *Ecological Economics*, in press (corrected proof, available online, accessed 4 February 2008).
Pacala, S. and R. Socolow (2004) 'Stabilisation wedges: solving the climate problem for the next 50 years with current technologies', *Science*, 305: 908–72.
Princen, T. (2005) *The Logic of Sufficiency*, Boston: MIT Press.
Raupach, M. R., G. Marland, P. Ciais, C. Le Quéré, J. G. C. Canadell, G. Klepper and C. B. Field (2007) 'Global and regional drivers of accelerating CO_2 emissions', *Proceedings of the National Academy of Sciences* (published online on 22 May 2007).
RCEP (2000) *Energy: the Changing Climate*, 22nd Report, Royal Commission on Environmental Pollution, London: The Stationery Office.
Roy, J. (2000) 'The rebound effect: some empirical evidence from India', *Energy Policy*, 28(6–7): 433–8.
Sanne, C. (2000) 'Dealing with environmental savings in a dynamical economy: how to stop chasing your tail in the pursuit of sustainability', *Energy Policy*, 28(6–7): 487–95.
Sanne, C. (2002) 'Willing consumers or locked in? Policies for sustainable consumption', *Ecological Economics*, 47: 273–87.
Sartori, I. and A. G. Hestnes (2007) 'Energy use in the life-cycle of conventional and low-energy buildings: a review article', *Energy and Buildings*, 39: 249–57.
Schipper, L. (1987) 'Energy conservation policies in the OECD: did they make a difference?' *Energy Policy*, 15: 538–48.
Schor, J. (1999) *The Overspent American: Upscaling, Downshifting, and the New Consumer*, New York: Basic Books.
Sorrell, S. (2007) *The Rebound Effect: an Assessment of the Evidence for Economy-wide Energy Savings from Improved Energy Efficiency*, London: UK Energy Research Centre.
Sorrell, S. and J. Dimitropoulos (2007a) *UKERC Review of Evidence for the Rebound Effect: Technical Report 3: Econometric Studies*, London: UK Energy Research Centre.
Sorrell, S. and J. Dimitropoulos (2007b) *UKERC Review of Evidence for the Rebound Effect: Technical Report 5: Energy, Productivity and Economic Growth Studies*, London: UK Energy Research Centre.
Sorrell, S. and J. Sijm (2003) 'Carbon trading in the policy mix', *Oxford Review of Economic Policy*, 19(3): 420–37.
Suh, S. (2006) 'Are services better for climate change?' *Environmental and Science and Technology*, 40(21): 6555–60.
WCED (1987) *Our Common Future*, World Commission on Environment and Development, Oxford: Oxford University Press.

Index